Sewage - Management and Treatment Techniques

Edited by Hassimi Abu Hasan

Published in London, United Kingdom

Sewage - Management and Treatment Techniques
http://dx.doi.org/10.5772/intechopen.1003552
Edited by Hassimi Abu Hasan

Contributors
Ali Shee, Cyrilus Otulo Wandera, Daniel Munga, Epere Aworabhi, Ernest Mbamalu Ezeh, Gorur Bettaiah Janardhana Swamy, Hassimi Abu Hasan, Hema Darshinni Vijayakumar, Huxley Mae Makonde, Jahira Alias, Jiawen Zhang, Laila Uweso Abubakar, Mariam Maku Swaleh, Mumini Ndenge Dzoga, Nur 'Izzati Ismail, Peter Chinedu Agu, Saeed Mutta Mwaguni, Setyo Budi Kurniawan, Toru Matsumoto

First published in London, United Kingdom, 2025 by IntechOpen
IntechOpen is the global imprint of INTECHOPEN LIMITED, registered in England and Wales, registration number: 11086078, 167-169 Great Portland Street, London, W1W 5PF, United Kingdom

For EU product safety concerns: IN TECH d.o.o., Prolaz Marije Krucifikse Kozulić 3, 51000 Rijeka, Croatia, info@intechopen.com or visit our website at intechopen.com.

British Library Cataloguing-in-Publication Data
A catalogue record for this book is available from the British Library

Sewage - Management and Treatment Techniques
Edited by Hassimi Abu Hasan
p. cm.
Print ISBN 978-1-83769-967-4
Online ISBN 978-1-83769-966-7
eBook (PDF) ISBN 978-1-83769-968-1

Meet the editor

Hassimi Abu Hasan is an Associate Professor in the Faculty of Engineering and Built Environment at the Universiti Kebangsaan Malaysia. He heads the Research Centre for Sustainable Process Technology (CESPRO). He is also a co-chair of the Environmental Engineering Education Specialized Group under the International Water Association. His expertise is developing innovative and cutting-edge methods for cost-effective treatment and resource recovery from water and wastewater using biofiltration, phytoremediation, and phycoremediation. He was awarded the National Young Engineering Icon Award 2024 by the Board of Engineers Malaysia (BEM) and the Malaysian Toray Science Foundation Award 2015. He has published over 200 scientific articles, 5 books, and 10 chapters in the book. He has collaborated with many nations, including the United Kingdom, China, India, Thailand, Indonesia, and Iraq.

Contents

Preface

I am grateful to God for granting me good health and a sharp mind and enabling the completion of this edited volume titled *Sewage – Management and Treatment Techniques*. This book comprises five chapters covering sewage treatment methods, resource recovery from sewage, sewage management through the Internet of Things, and the socio-economic and environmental impacts. The book has been produced with the highest quality and contains the latest information. It serves as a valuable reference for those involved in the sewage management and treatment sector. Additionally, it appeals to a wide range of readers, including policymakers, academicians, researchers, engineers, and students.

I hope the insights shared in this book will encourage all stakeholders to manage and treat sewage efficiently. I would like to express my sincere gratitude to the Ministry of Higher Education, Malaysia, for funding this research through the Transdisciplinary Research Grant Scheme (TRGS) under the project titled Resilience and Security of Water through Sustainable Localised Domestic Wastewater Reclamation for Landscaping and Toilet Use (Grant No.: TRGS/1/2022/UKM/02/3). My heartfelt thanks also go to the Universiti Kebangsaan Malaysia for continuous support and provision of various facilities, which have contributed significantly to the dissemination of knowledge through the publication of this book.

I also extend my appreciation to the editors at IntechOpen for their invaluable support throughout the publication process, from the initial book proposal to editing and final publication. Lastly, my deepest gratitude goes to my family for their unwavering support, allowing me to continue contributing my expertise toward human capital development and societal progress.

Hassimi Abu Hasan
Faculty of Engineering and Built Environment,
Department of Chemical and Process Engineering,
Universiti Kebangsaan Malaysia,
Selangor, Malaysia

Chapter 1

Biological Treatment Techniques for Sewage: Aerobic and Anaerobic Processes

Ernest Mbamalu Ezeh, Peter Chinedu Agu and Epere Aworabhi

Abstract

Sewage treatment is crucial to safeguard public health and the environment. Various techniques are available for treating wastewater, and the selection of one often depends on factors like the wastewater composition, environmental conditions, and treatment objectives. One of the most widely used methods of biological treatment for sewage is activated sludge, which employs aerobic bacteria to degrade organic matter. This process involves blending wastewater with a mixture of bacteria and flocs, with the bacteria consuming the organic matter to produce clear water while the flocs settle. Another commonly employed method is anaerobic digestion, which employs anaerobic bacteria to decompose organic matter in the absence of oxygen. These methods generate biogas, containing methane and carbon dioxide, and digestate, a nutrient-rich sludge suitable as soil amendment or fertilizer. This chapter explores the mechanisms and applications of activated sludge and anaerobic digestion in biological sewage treatment, highlighting their advantages and downsides, the factors that influence their performance, such as pH, temperature, and substrate concentration. It covers the latest improvements in biological sewage treatment, including the use of new microbial strains and hybrid treatment systems.

Keywords: sewage management, biological sewage treatment, microorganisms, biogas, digestate

1. Introduction

The treatment of sewage is a crucial aspect of modern public health and environmental protection. Effective sewage management is essential to prevent the spread of diseases and to protect water resources, which are vital for human consumption, agriculture, and industrial use. The evolution of sewage management has been driven by the need to mitigate the adverse effects of untreated wastewater on ecosystems and human communities [1, 2]. Untreated sewage contains a variety of contaminants, including pathogens, organic matter, nutrients, and toxic substances, which can lead to severe environmental degradation and health issues if released into natural water bodies [3].

Historical developments highlight the transition from rudimentary disposal methods to sophisticated treatment processes, underscoring the growing understanding of sewage's impact on water bodies and public health [4]. In ancient times, sewage was often disposed of in nearby water bodies without any treatment, leading to severe pollution and health problems [5, 6]. The Industrial Revolution marked a significant turning point, as the increased urbanization and industrial activities resulted in larger volumes of wastewater, necessitating more systematic treatment approaches [6]. Early methods focused primarily on physical processes such as sedimentation and filtration [2]. However, it became evident that these methods were insufficient to remove dissolved organic matter and pathogens [5]. The development of biological treatment processes in the late nineteenth and early twentieth centuries represented a major advancement, as these processes could effectively reduce organic pollutants through the activity of microorganisms [6].

Biological treatment processes, which harness the natural activities of microorganisms to degrade organic pollutants, have become fundamental in modern sewage treatment systems. These processes are categorized broadly into aerobic and anaerobic treatments, each with distinct mechanisms and applications [7]. The literature indicates that aerobic treatment, particularly the activated sludge process, has been extensively adopted due to its effectiveness in reducing organic matter and its adaptability to varying wastewater compositions [8, 9]. The activated sludge process operates on the principle of maintaining a suspended growth system where aerobic bacteria consume organic pollutants, forming flocs that can be easily separated from the treated water [10]. The activated sludge process involves several stages, including aeration, sedimentation, and sludge recycling [11]. During the aeration stage, wastewater is mixed with a microbial culture in the presence of oxygen. Aerobic bacteria metabolize the organic matter, converting it into carbon dioxide, water, and biomass. The resulting microbial flocs are then separated from the treated water in the sedimentation stage [6]. Some of the settled sludge is recycled back to the aeration tank to maintain a high concentration of microorganisms, while the excess sludge is removed for further treatment or disposal. This process not only reduces organic pollutants but also helps in the removal of nitrogen and phosphorus, which are important for preventing eutrophication in receiving waters [9].

Anaerobic digestion, another widely researched and applied biological treatment method, leverages anaerobic bacteria to decompose organic matter in the absence of oxygen [12]. This process not only reduces the volume of waste but also produces biogas, a valuable by-product that can be utilized for energy generation [6]. The anaerobic digestion process occurs in several stages, including hydrolysis, acidogenesis, acetogenesis, and methanogenesis. During hydrolysis, complex organic molecules are broken down into simpler compounds [13]. Acidogenesis and acetogenesis further convert these compounds into volatile fatty acids, hydrogen, and carbon dioxide. Finally, during methanogenesis, methanogenic bacteria produce methane and carbon dioxide from these intermediates. The digestate resulting from anaerobic digestion is nutrient-rich and can be used as fertilizer or soil amendment, adding an economic benefit to the environmental advantages. Anaerobic digestion is particularly suited for treating high-strength wastewater and sludge due to its ability to handle high organic loads and produce valuable by-products. Additionally, it generates less excess sludge compared to aerobic processes, reducing the need for sludge handling and disposal [14].

The performance of these biological treatment processes is influenced by various factors, including temperature, pH, and substrate concentration. Research has shown that optimizing these parameters can significantly enhance treatment efficiency and stability [15, 16]. For instance, temperature affects the metabolic rates of microorganisms, with mesophilic and thermophilic conditions being optimal for different types of bacteria. pH levels can influence microbial activity and the solubility of nutrients and toxins. Substrate concentration, or the amount of organic matter available for microbial consumption, must be balanced to prevent overloading the system and to ensure efficient degradation [17]. Recent advancements in the field have focused on the development of new microbial strains with enhanced degradation capabilities and the integration of hybrid systems that combine aerobic and anaerobic processes to capitalize on the strengths of both methods [15]. For example, genetically engineered microorganisms and microbial consortia have been developed to degrade specific pollutants more efficiently. Hybrid systems, such as sequential batch reactors and membrane bioreactors, integrate aerobic and anaerobic stages to achieve higher treatment efficiencies and resource recovery [17]. Therefore, this chapter aims to provide a comprehensive overview of the mechanisms, applications, and performance factors of activated sludge and anaerobic digestion processes. These innovations aim to address the limitations of conventional treatment methods and improve the sustainability of sewage treatment practices.

2. Basics of biological sewage treatment

2.1 Definition and principles

Biological sewage treatment refers to the use of microorganisms to remove contaminants, primarily organic matter from wastewater [18]. This process relies on the metabolic activities of various microorganisms to degrade and convert pollutants into stable forms [19]. The principles underlying biological sewage treatment include:

a. Biochemical oxygen demand (BOD) reduction: Microorganisms consume organic matter in the wastewater, thereby reducing the BOD, which is a measure of the amount of oxygen required to biologically decompose the organic material present in the water [20].

b. Microbial metabolism: Microorganisms metabolize organic pollutants through aerobic or anaerobic processes, breaking them down into simpler, nontoxic compounds such as carbon dioxide, water, methane, and biomass [20].

c. Flocculation and sedimentation: In aerobic processes like activated sludge, microorganisms form flocs that aggregate and settle, allowing for the separation of treated water and microbial biomass [12].

2.2 Role of microorganisms in wastewater treatment

Microorganisms play a pivotal role in the biological treatment of sewage by facilitating the breakdown of organic matter and other pollutants [21].

a. Organic matter degradation: Bacteria, fungi, and protozoa consume organic pollutants, converting them into simpler compounds. Aerobic bacteria require oxygen to perform this process, while anaerobic bacteria thrive in oxygen-free environments.

b. Nutrient removal: Certain bacteria are capable of removing nutrients such as nitrogen and phosphorus from wastewater. Nitrifying bacteria convert ammonia into nitrate, which can then be further processed by denitrifying bacteria to produce nitrogen gas, reducing nutrient loads in the effluent.

c. Pathogen reduction: Some microorganisms can outcompete or consume pathogenic bacteria and viruses, contributing to the overall reduction of harmful pathogens in treated wastewater.

d. Biogas production: In anaerobic digestion, microorganisms decompose organic matter to produce biogas, primarily composed of methane and carbon dioxide, which can be captured and used as a renewable energy source.

e. Biofilm formation: In certain treatment systems like trickling filters and bio-towers, microorganisms form biofilms on the surfaces of the media. These biofilms facilitate the treatment process by providing a large surface area for microbial activity and organic matter degradation.

Comprehending the underlying principles and the pivotal functions of microorganisms is imperative for the efficient optimization and administration of biological sewage treatment procedures. The effective and ecologically sustainable design and operation of treatment systems are supported by this fundamental understanding.

3. Activated sludge process (aerobic treatment)

3.1 Overview of the activated sludge process

The activated sludge process is a widely used method in wastewater treatment that involves the use of aerobic microorganisms to decompose organic pollutants in sewage. In this process, wastewater is mixed with a microbial culture, known as activated sludge, in the presence of oxygen [22]. The microorganisms metabolize the organic matter, forming aggregates called flocs, which can be easily separated from the treated water [9]. Historically, the activated sludge process was developed in the early twentieth century by Edward Ardern and William Lockett in England. Their pioneering work in 1914 demonstrated that aerating sewage with microorganisms could effectively reduce organic pollutants [23]. Since then, the process has evolved and been widely adopted due to its efficiency and adaptability to various types of wastewaters [20].

3.2 Mechanisms of activated sludge

3.2.1 Microbial activity and organic matter degradation

The activated sludge process is a cornerstone of modern wastewater treatment, primarily driven by the metabolic activities of a diverse microbial consortium. The degradation of organic matter in this system involves a series of complex biochemical and microbial interactions, with aerobic bacteria playing a central role [24]. The microbial community within activated sludge is highly diverse, comprising bacteria, protozoa, fungi, and archaea [25]. Bacteria, particularly aerobic heterotrophs, are the most significant contributors to organic matter degradation [26]. Key bacterial genera include *Pseudomonas, Bacillus, Nitrosomonas*, and *Nitrobacter*, each participating in different stages of organic compound breakdown [27]. The process of organic matter degradation involves several interrelated biochemical pathways [28]. Initially, large organic polymers such as proteins, lipids, and polysaccharides are hydrolyzed by extracellular enzymes (proteases, lipases, and cellulases) into smaller monomers (amino acids, fatty acids, and sugars) [29, 30]. This crucial initial step renders the organic matter accessible to microbial uptake. Aerobic bacteria utilize oxygen as a terminal electron acceptor in their metabolic processes [31]. The monomers produced during hydrolysis are further oxidized through glycolysis, the tricarboxylic acid cycle, and oxidative phosphorylation, leading to the production of carbon dioxide, water, and energy in the form of adenosine triphosphate (ATP) [31, 32]. This energy is essential for microbial growth and maintenance [33]. Additionally, a fraction of the degraded organic matter is assimilated into new microbial biomass, which is critical for the sustenance and proliferation of the microbial community [33].

Several environmental parameters critically influence microbial activity and organic matter degradation efficiency in the activated sludge process. Sufficient dissolved oxygen (DO) levels are imperative for aerobic respiration [34]. Low DO can lead to the proliferation of facultative anaerobes and the onset of incomplete degradation pathways, resulting in suboptimal treatment performance [35]. The sludge retention time (SRT) or mean cell residence time (MCRT) dictates the time microorganisms remain in the system [36]. Optimal SRT ensures the retention of slow-growing but essential microbial species, enhancing overall degradation efficiency. Additionally, the enzymatic activity and microbial growth rates are pH and temperature-dependent, with optimal conditions typically ranging between pH 6.5–8.5 [37] and temperatures of 20–35°C [38], although specific microbes may thrive under different conditions.

To optimize organic matter degradation, several advanced strategies and technologies have been employed. Bioaugmentation involves the introduction of specialized microbial strains to bolster the degradation of specific contaminants or to enhance overall process stability [39]. Membrane bioreactors (MBRs) combine activated sludge with membrane filtration technology, allowing for higher biomass concentrations and improved effluent quality [40]. The implementation of real-time monitoring systems and automated control strategies helps maintain optimal environmental conditions and promptly address any process perturbations.

3.2.2 Formation and role of flocs

As the microorganisms metabolize the organic matter, they produce extracellular polymeric substances (EPS), which help in the aggregation of cells into flocs [41]. These flocs are dense clusters of bacteria, protozoa, and other microorganisms. The flocs have a higher settling rate, allowing them to be easily separated from the treated water during the sedimentation stage. The settled flocs, known as sludge, can be recycled back into the aeration tank to maintain a high concentration of microorganisms or removed for further treatment [15].

3.3 Process configuration

3.3.1 Components of activated sludge systems

An activated sludge system typically comprises several key components essential for the efficient treatment of wastewater. As shown in **Table 1**, these components

Components	Description
Aeration tank	Where wastewater is mixed with activated sludge and aerated to provide oxygen for microbial activity.
Sedimentation tank (Clarifier)	Where the flocs settle, separating the treated water from the sludge.
Sludge recycling system	Which returns a portion of the settled sludge to the aeration tank to maintain microbial concentration.
Sludge disposal system	For handling excess sludge that is not recycled.

Table 1.
Description of components of activated sludge system.

include the aeration tank, where wastewater is mixed with microbial biomass and aerated to promote the breakdown of organic matter. Following this, the mixture flows into a sedimentation tank, also known as a secondary clarifier, where the activated sludge settles out, separating from the treated effluent [39].

3.3.2 Process flow and stages

The operations of the flow process include pretreatment, aeration, sedimentation, disinfection, and sludge handling (**Table 2**). The activated sludge system is a widely used process in wastewater treatment, involving several key stages and operations. It begins with the primary treatment, where large solids are removed from the wastewater. The primary effluent then enters an aeration tank, where it is mixed with a microbial biomass called activated sludge. Oxygen is supplied to the tank through aerators to support the aerobic microorganisms that decompose organic matter [29]. The mixture then flows to a secondary clarifier, where the activated sludge settles out, and the treated effluent is separated and discharged. A portion of the settled sludge, known as return-activated sludge, is recirculated back to the aeration tank to maintain the microbial population. The excess sludge, or waste-activated sludge, is removed for further treatment or disposal. This continuous cycle ensures the efficient breakdown of organic pollutants and the production of clean effluent [15].

3.4 Benefits and limitations

The sludge system offers several advantages. It is effective in removing organic matter, nutrients, and pathogens from wastewater, resulting in relatively high-quality

Process	Operations
Pretreatment	Removal of large solids and grit to protect downstream processes.
Aeration	Mixing wastewater with activated sludge and supplying oxygen.
Sedimentation	Settling of flocs in a clarifier to separate treated water from sludge.
Disinfection	This is an optional further treatment to kill the remaining pathogens.
Sludge handling	Recycling and disposal of settled sludge.

Table 2.
Flow process and operations of activated sludge system.

effluent [35]. The process is also adaptable to various scales, from small community systems to large municipal plants [37]. Furthermore, it produces biogas during anaerobic digestion, which can be used as a renewable energy source, contributing to energy savings and sustainability. Additionally, the treated sludge can be utilized as a fertilizer or soil conditioner, promoting resource recovery and reducing waste disposal needs [38].

However, the sludge system also has notable disadvantages. The initial setup and maintenance costs can be high, requiring significant investment in infrastructure and skilled personnel. The process generates large volumes of sludge, which necessitates proper handling, treatment, and disposal, posing environmental and logistical challenges. There is also a risk of odor and potential health hazards if the sludge is not adequately managed [20]. Moreover, the system can be energy-intensive, particularly in the aeration stages, leading to increased operational costs and carbon footprint [12]. Lastly, regulatory compliance and monitoring requirements can be stringent, adding to the complexity and cost of managing sludge systems effectively.

4. Anaerobic digestion process

4.1 Overview of anaerobic digestion

Anaerobic digestion is a biological process in which microorganisms break down organic matter in the absence of oxygen. This process results in the production of biogas, primarily composed of methane and carbon dioxide, and a nutrient-rich digestate. Anaerobic digestion is commonly used for treating high-strength wastewater and organic solid wastes, including sewage sludge, agricultural residues, and industrial effluents [20]. Historically, the use of anaerobic digestion dates back to ancient times, with early civilizations employing rudimentary methods to treat organic waste. The modern development of anaerobic digestion began in the nineteenth century, with significant advancements in understanding microbial processes and reactor designs. The first full-scale anaerobic digester was built in 1859 in Bombay, India. Since then, the technology has evolved, with widespread applications in wastewater treatment, renewable energy production, and waste management [4].

4.2 Mechanisms of anaerobic digestion

4.2.1 Microbial activity and organic matter degradation

Anaerobic digestion involves a series of microbial processes that convert complex organic matter into simpler compounds [12, 42]. The process occurs in four main stages shown in **Figure 1**

a. Hydrolysis: Complex organic molecules such as carbohydrates, proteins, and fats are broken down into simpler sugars, amino acids, and fatty acids.

b. Acidogenesis: The products of hydrolysis are further converted into volatile fatty acids, alcohols, hydrogen, and carbon dioxide by acidogenic bacteria.

c. Acetogenesis: Acidogenic intermediates are converted into acetic acid, hydrogen, and carbon dioxide by acetogenic bacteria.

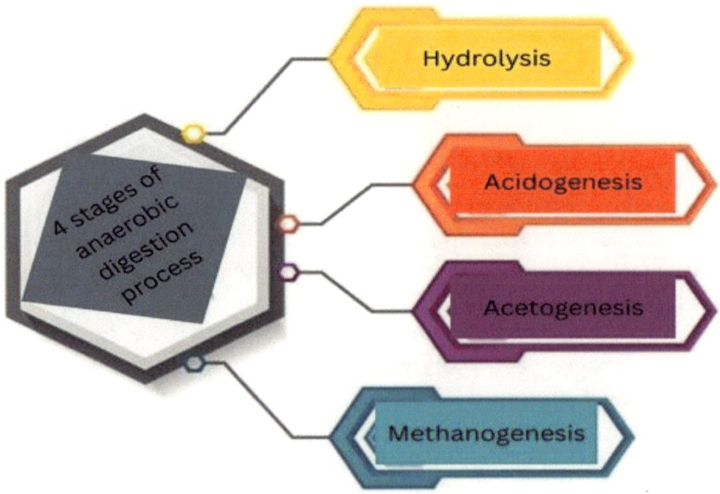

Figure 1.
Anaerobic processes in organic matter degradation.

d. Methanogenesis: Methanogenic archaea produce methane and carbon dioxide from acetic acid, hydrogen, and carbon dioxide.

4.2.2 Production of biogas

Biogas is a valuable by-product of anaerobic digestion, consisting mainly of methane (CH_4) and carbon dioxide (CO_2), with trace amounts of other gases such as hydrogen sulfide (H_2S). Methane, a potent greenhouse gas, can be captured and utilized as a renewable energy source for electricity generation, heating, or as a vehicle fuel. The production of biogas not only provides an alternative energy source but also helps reduce greenhouse gas emissions [43, 44].

4.3 Process configuration

4.3.1 Components and stages of anaerobic digestion systems

As shown in **Table 3**, aerobic digestion systems involve the breakdown of organic matter in the presence of oxygen by aerobic microorganisms. These systems typically include components such as aeration tanks, where oxygen is supplied to the microbial community to facilitate the decomposition process, and secondary clarifiers, where solids are separated from the treated effluent. The process includes stages like initial aeration, where organic matter is rapidly broken down, followed by stabilization, where further decomposition reduces the volume and mass of the sludge. The end products of aerobic digestion are carbon dioxide, water, and a stabilized sludge that can be used as a soil amendment or disposed of safely [43].

4.4 Operational parameters

Figure 2 highlights several factors that influence the efficiency of anaerobic digestion.

Components	Description
a. Digester Tank	A sealed, oxygen-free container where the anaerobic digestion process occurs.
b. Feedstock Storage and Preparation	Facilities for storing and preparing organic waste materials before feeding them into the digester.
c. Biogas Collection and Storage	Systems for capturing and storing the produced biogas.
d. Digestate Storage and Handling	Facilities for handling the nutrient-rich digestate, which can be used as fertilizer or soil amendment.
Flow process stages	**Operations**
a. Feedstock Preparation	Organic waste is collected, sorted, and pre-treated to enhance its biodegradability.
b. Digestion	The feedstock is deposited in a digester tank, where anaerobic microorganisms break down the organic matter over several weeks.
c. Biogas Collection	The produced biogas is collected from the top of the digester tank and stored for further use.
d. Digestate Handling	The remaining digestate is removed from the digester, stored, and eventually used as a fertilizer or soil conditioner.

Table 3.
Description of aerobic digestion system.

Figure 2.
Factors that influence anaerobic digestion.

Temperature is a critical factor influencing the rate and efficiency of anaerobic digestion, which can occur under three temperature ranges: psychrophilic (below 20°C, with a slow digestion rate), mesophilic (20–45°C, the optimal range for most digesters, typically around 35°C), and thermophilic (45–60°C, with a higher digestion

Benefits	Limitations and Challenges
Renewable Energy Production: Generates biogas that can be used as a sustainable energy source.	High Capital Costs: Initial investment for anaerobic digestion systems can be substantial.
Reduction of Waste Volume: Significantly reduces the volume of organic waste, minimizing disposal requirements. **Nutrient Recovery**: Produces nutrient-rich digestate that can be used as fertilizer, enhancing soil health and agricultural productivity.	Technical Complexity: Requires careful monitoring and control of operational parameters to ensure stable performance. Sensitivity to Environmental Conditions: Microbial activity can be affected by fluctuations in temperature, pH, and substrate composition.
Greenhouse Gas Mitigation: Captures methane that would otherwise be released into the atmosphere, reducing greenhouse gas emissions.	Potential for Toxic Compounds: Certain feedstocks may contain substances that inhibit microbial activity, requiring pretreatment or careful management.
Odor Control: Reduces odors associated with organic waste decomposition.	

Table 4.
Benefits and limitations of anaerobic digestion.

rate but more energy-intensive) [45]. The pH of the digestion environment affects microbial activity and process stability, with an optimal range between 6.5 and 8.0, and maintaining pH within this range is essential to prevent the inhibition of microbial processes [46]. The concentration of organic matter (substrate) in the feedstock impacts the digestion process, where high substrate concentrations can lead to rapid biogas production but may also cause system overload and process instability, thus balancing substrate concentration is crucial for maintaining stable operation [47]. Retention time, which refers to the duration that the feedstock remains in the digester, ensures complete digestion of organic matter and optimal biogas production, with typical retention times ranging from 15 to 30 days, depending on the feedstock and operating conditions [48].

4.5 Benefits and limitations of anaerobic digestion

Anaerobic digestion offers several benefits, including the production of biogas, a renewable energy source that can be used for electricity and heat, thus reducing reliance on fossil fuels. It also produces a nutrient-rich digestate that can be used as a fertilizer, enhancing soil health [45]. Additionally, anaerobic digestion helps in reducing greenhouse gas emissions and managing organic waste more effectively. However, it has limitations, such as the need for careful monitoring and control of the digestion process, high initial setup costs, and the potential for odor issues [46]. Furthermore, the efficiency of the process can be affected by the variability in feedstock composition and the need for pretreatment in some cases (see **Table 4**).

From the highlights of the benefits and limitations, notably, optimizing anaerobic digestion for effective waste treatment and resource recovery requires a thorough understanding of its principles and operating parameters.

5. Comparative analysis of aerobic and anaerobic processes

The distinct properties of the aerobic and anaerobic processes are succinctly highlighted.

5.1 Efficiency in organic matter removal

In aerobic processes, activated sludge processes, like those used in wastewater treatment, can efficiently remove up to 95% of Biochemical Oxygen Demand (BOD) due to the rapid metabolism of aerobic microorganisms. However, anaerobic digestion effectively reduces organic matter in wastewater, removing 50–70% of chemical oxygen demand (COD), but is slower due to lower metabolic rates of microorganisms.

5.2 Energy requirements and production

Large-scale systems require significant energy input for aeration to maintain oxygen levels for aerobic microorganisms, which can be a significant operational cost. However, anaerobic digestion, a process without aeration, offers lower energy requirements and biogas production, which can be used as a renewable energy source or surplus energy [49].

5.3 Environmental impact

Aerobic processes produce less odor and reduce organic pollutants, but their high energy demand, often from nonrenewable sources, can increase their carbon footprint. Biogas production reduces fossil fuel reliance and promotes renewable energy. Digestate can be used as fertilizer, promoting nutrient recycling. Drawbacks include methane emissions and odor issues [50].

5.4 Application scenarios

For aerobic processes, the high efficiency wastewater treatment system is ideal for municipal applications requiring quick processing times and low energy costs, aiming to meet stringent discharge standards and improve water quality [51]. However, in anaerobic, the product is ideal for high-strength industrial wastewater, agricultural wastes, and sewage sludge with high organic loads, particularly in scenarios requiring energy recovery and waste stabilization, especially in rural or decentralized locations [52].

As comparatively outlined, selecting the best treatment option depends on several factors, including the characteristics of the wastewater, operational objectives, and environmental concerns. It is helpful to understand the relative benefits and drawbacks of aerobic and anaerobic processes.

6. Factors influencing performance

6.1 Environmental conditions

6.1.1 Temperature

For Aerobic Processes, optimal microbial activity typically occurs within the mesophilic range (15–35°C). Higher temperatures can enhance reaction rates but may lead to increased energy consumption for aeration. Conversely, lower temperatures can slow down microbial metabolism, reducing treatment efficiency. However, Anaerobic digestion can operate under psychrophilic (below 20°C), mesophilic (20–45°C), and thermophilic (45–60°C) conditions. Mesophilic and thermophilic conditions are preferred for their higher biogas production rates and better pathogen reduction, though thermophilic digestion requires more energy input for heating [45].

6.1.2 pH levels

For Aerobic Processes, the optimal pH for aerobic treatment is generally between 6.5 and 8.5. Deviation from this range can inhibit microbial activity and reduce the efficiency of organic matter degradation [53]. However, anaerobic digestion performs best within a pH range of 6.5–8.0. Maintaining a stable pH is critical, as acidic conditions (pH below 6.5) can inhibit methanogenic bacteria, while alkaline conditions (pH above 8.0) can negatively affect other microbial communities involved in the digestion process [54].

6.2 Wastewater composition

6.2.1 Organic load

For Aerobic Processes, high organic loads can enhance microbial growth and activity but may also lead to issues like sludge bulking and poor settling if not properly managed. Balancing the organic load is crucial for maintaining system stability and efficiency. Similarly, Anaerobic digestion is particularly effective for high-strength wastewaters with high organic loads. However, excessive organic loading can lead to process instability, accumulation of volatile fatty acids, and inhibition of methanogenesis [53].

6.2.2 Presence of inhibitors

For Aerobic Processes, toxic compounds, such as heavy metals, pesticides, and industrial chemicals, can inhibit microbial activity and reduce treatment efficiency. Pretreatment of wastewater to remove or neutralize inhibitors may be necessary [55]. Similarly, Anaerobic microorganisms are sensitive to inhibitors like ammonia, sulfides, and certain organic compounds. High concentrations of these substances can disrupt microbial communities and biogas production. Monitoring and managing inhibitor levels are essential for stable operation [56].

6.3 System design and operation

6.3.1 Reactor design

In Aerobic Processes, common reactor designs include completely mixed aeration tanks, plug flow reactors, and sequencing batch reactors. The choice of design influences oxygen transfer efficiency, mixing, and overall treatment performance. Proper design ensures effective contact between microorganisms and organic matter [45]. However, Anaerobic digesters come in various designs, such as continuously stirred tank reactors (CSTRs), up-flow anaerobic sludge blanket (UASB) reactors, and anaerobic sequencing batch reactors. Design considerations include retention time, mixing, and biogas collection efficiency to optimize digestion performance [55].

6.3.2 Operational practices

Aerobic Processes' effective operational practices include maintaining appropriate dissolved oxygen levels, regular monitoring of sludge age and concentration, and periodic removal of excess sludge [28]. Consistent monitoring and control of operational parameters ensure optimal microbial activity and treatment efficiency. However, anaerobic processes' key operational practices involve maintaining stable temperature and pH, regular feeding of substrates, and effective management of biogas production and digestate disposal [57]. Monitoring parameters like volatile fatty acid concentration, alkalinity, and biogas composition helps in maintaining stable digestion processes.

To ensure effective pollutant removal, achieve stable and sustainable operation, and maximize the performance of both aerobic and anaerobic sewage treatment processes, it is imperative to comprehend and manage these aspects.

7. Recent advancements in biological sewage treatment

7.1 New microbial strains

7.1.1 Development and application

a. Genetic engineering: Advances in genetic engineering have enabled the development of microbial strains with enhanced capabilities for degrading specific pollutants, increasing treatment efficiency. These genetically modified microorganisms can target hard-to-degrade compounds such as pharmaceuticals and industrial chemicals, offering improved performance in both aerobic and anaerobic processes [58, 59].

b. Bioaugmentation: The introduction of specialized microbial strains to wastewater treatment systems, known as bioaugmentation, can enhance the breakdown of specific contaminants and improve overall system performance. This approach has been particularly effective in treating industrial wastewater with high concentrations of recalcitrant compounds [39].

c. Consortia of microbes: Research has focused on creating microbial consortia, where multiple strains work synergistically to degrade complex organic matter more efficiently [60]. These consortia are designed to exploit the complementary metabolic pathways of different microorganisms, resulting in more robust and resilient treatment processes [24].

d. Adaptation to extreme conditions: Microbial strains that can thrive under extreme conditions, such as high salinity, low pH, or high temperatures, have been developed [60]. These strains expand the applicability of biological treatment to a wider range of wastewater, including those from specific industrial sectors or regions with harsh climates [61].

7.2 Hybrid treatment systems

7.2.1 Combination of aerobic and anaerobic processes

a. Integrated systems: Hybrid treatment systems combine aerobic and anaerobic processes to leverage the advantages of both methods. For instance, an anaerobic digester can first reduce the organic load and produce biogas, followed by an aerobic treatment to polish the effluent and achieve higher water quality standards [62].

b. Sequential reactors: In these systems, wastewater undergoes sequential treatment in separate anaerobic and aerobic reactors. This configuration allows for the phased removal of different types of pollutants, optimizing overall treatment efficiency [63].

c. Simultaneous processes: Some hybrid systems integrate both aerobic and anaerobic processes within a single reactor, creating distinct zones for each process. This approach can enhance nutrient removal and reduce the footprint of treatment facilities [64].

7.3 Benefits and challenges

7.3.1 Benefits

a. Enhanced treatment efficiency: Combining aerobic and anaerobic processes can achieve higher overall pollutant removal rates, including more effective degradation of complex organic compounds and nutrients.

b. Energy optimization: Anaerobic digestion can reduce the organic load and produce biogas, which can be used to power aerobic treatment systems, offsetting energy costs.

c. Improved sludge management: Hybrid systems can reduce the volume and improve the stability of sludge, making it easier to handle and dispose of.

d. Flexibility and resilience: The integration of both processes can provide greater operational flexibility and resilience to fluctuating wastewater characteristics and loads.

7.3.2 Challenges

a. The complexity of design and operation: Hybrid systems require careful design and precise control of operational parameters to ensure that both aerobic and anaerobic processes function optimally. This complexity can increase capital and operational costs.

b. Integration issues: Ensuring seamless integration between aerobic and anaerobic stages can be challenging, particularly in maintaining appropriate environmental conditions for each process.

c. Balancing microbial communities: Maintaining balanced and healthy microbial communities in both aerobic and anaerobic zones is critical for consistent performance. Disruptions in one part of the system can impact the overall treatment efficiency.

d. Monitoring and control: Hybrid systems require advanced monitoring and control systems to manage the interactions between aerobic and anaerobic processes, detect potential issues, and optimize performance.

Therefore, the efficiency, adaptability, and sustainability of biological sewage treatment have been significantly improved by recent developments in microbial technology and hybrid treatment systems. In a world that is changing quickly, these technologies have the potential to address the mounting problems associated with wastewater management.

8. Conclusion

Biological treatment techniques play a crucial role in sewage management, focusing on the core principles, mechanisms, and applications of aerobic and anaerobic processes. The activated sludge process is noted for its high efficiency and adaptability, while anaerobic digestion is recognized for its energy recovery and waste volume reduction benefits. Both methods have their distinct advantages and face unique challenges, influenced significantly by factors such as environmental conditions, wastewater composition, and system design. Recent advancements, such as the development of new microbial strains and hybrid treatment systems, have further enhanced the effectiveness and sustainability of biological sewage treatment. These innovations are essential for addressing the evolving challenges in wastewater management, ultimately improving public health and environmental protection.

Acknowledgements

We are grateful for the administrative support offered to us during this study by the Department of Chemical Engineering, Faculty of Engineering, Federal University, Otuoke, Bayelsa State, Nigeria. The authors acknowledge the use of ChatGPT for language editing in the chapter.

Conflict of interest

The authors declare no conflict of interest.

Author details

Ernest Mbamalu Ezeh[1*], Peter Chinedu Agu[2] and Epere Aworabhi[1]

1 Faculty of Engineering, Department of Chemical Engineering, Federal University, Otuoke, Nigeria

2 Department of Biochemistry, College of Science, Evangel University, Akaeze, Nigeria

*Address all correspondence to: ezehem@fuotuoke.edu.ng

References

[1] Rorat A, Courtois P, Vandenbulcke F, Lemiere S. Sanitary and environmental aspects of sewage sludge management. In: Industrial and Municipal Sludge. France: University of Lille, Elsevier Science; 2019. pp. 155-180. DOI: 10.1016/B978-0-12-815907-1.00008-8

[2] Christodoulou A, Stamatelatou K. Overview of legislation on sewage sludge management in developed countries worldwide. Water Science and Technology. 2016;**73**:453-462

[3] Connor R, Renata A, Ortigara C, Koncagül E, Uhlenbrook S, Lamizana-Diallo BM, et al. The United Nations world water development report 2017. Wastewater: The untapped resource. The United Nations World Water Development Report. Catalysis Today. 2017;**147**(1):1-59

[4] Fijalkowski K, Rorat A, Grobelak A, Kacprzak MJ. The presence of contaminations in sewage sludge—The current situation. Journal of Environmental Management. 2017;**203**:1126-1136

[5] Yadav B, Pandey AK, Kumar LR, Kaur R, Yellapu SK, Sellamuthu B, et al. Introduction to wastewater microbiology: Special emphasis on hospital wastewater. In: Current Developments in Biotechnology and Bioengineering. Cambridge, United States: Elsevier; 2020. pp. 1-41. DOI: 10.1016/B978-0-12-819722-6.00001-8

[6] Bradu P, Biswas A, Nair C, Sreevalsakumar S, Patil M, Kannampuzha S, et al. Recent advances in green technology and industrial revolution 4.0 for a sustainable future. Environmental Science and Pollution Research International. 2023;**30**(60):124488-124519. DOI: 10.1007/s11356-022-20024-4

[7] Sravan JS, Matsakas L, Sarkar O. Advances in biological wastewater treatment processes: Focus on low-carbon energy and resource recovery in biorefinery context. Bioengineering (Basel). 2024;**11**(3):281. DOI: 10.3390/bioengineering11030281

[8] Roy H, Rahman TU, Tasnim N, Arju J, Rafid MM, Islam MR, et al. Microbial fuel cell construction features and application for sustainable wastewater treatment. Membranes (Basel). 2023;**13**(5):490. DOI: 10.3390/membranes13050490

[9] Modin O, Persson F, Wilén BM, Hermansson M. Nonoxidative removal of organics in the activated sludge process. Critical Reviews in Environmental Science and Technology. 2016;**46**(7):635-672. DOI: 10.1080/10643389.2016.1149903

[10] Tchobanoglous G, Burton FL, Stensel HD, Metcalf & Eddy Inc. Wastewater Engineering, Treatment and Reuse. 4th ed. New York, USA: McGraw-Hill; 2004

[11] Leu S-Y, Chan L, Stenstrom MK. Toward long solids retention time of activated sludge processes: Benefits in energy saving, effluent quality, and stability. Water Environment Research. 2012;**84**:42-53

[12] Harirchi S, Wainaina S, Sar T, Nojoumi SA, Parchami M, Parchami M, et al. Microbiological insights into anaerobic digestion for biogas, hydrogen or volatile fatty acids (VFAs): A review. Bioengineered. 2022;**13**(3):6521-6557. DOI: 10.1080/21655979.2022.2035986

[13] Sekoai PT, Ghimire A, Ezeokoli OT, Rao S, Ngan WY, Habimana O, et al. Valorization of volatile fatty acids from the dark fermentation waste streams-a promising pathway for a biorefinery concept. Renewable and Sustainable Energy Reviews. 2021;**143**:110971

[14] Chew KR, Leong HY, Khoo KS, Vo DVN, Anjum H, Chang CK, et al. Effects of anaerobic digestion of food waste on biogas production and environmental impacts: A review. Environmental Chemistry Letters. 2021;**19**(4):2921-2939

[15] Khalidi-Idrissi A, Madinzi A, Anouzla A, Pala A, Mouhir L, Kadmi Y, et al. Recent advances in the biological treatment of wastewater rich in emerging pollutants produced by pharmaceutical industrial discharges. International Journal of Environmental Science and Technology (Tehran). 2023;**20**:11719-11740. DOI: 10.1007/s13762-023-04867-z

[16] Peña G, Israel O, Zavala MÁL, Ruelas HC. Pharmaceuticals market, consumption trends, and disease incidence are not driving the pharmaceutical research on water and wastewater. International Journal of Environmental Research and Public Health. 2021;**18**(5):2532

[17] Gonzalez JM, Aranda B. Microbial growth under limiting conditions-future perspectives. Microorganisms. 2023;**11**(7):1641. DOI: 10.3390/microorganisms11071641

[18] Bala S, Garg D, Thirumalesh BV, Sharma M, Sridhar K, Inbaraj BS, et al. Recent strategies for bioremediation of emerging pollutants: A review for a green and sustainable environment. Toxics. 2022;**10**(8):484. DOI: 10.3390/toxics10080484

[19] Masindi V, Osman MS, Tekere M. Mechanisms and approaches for the removal of heavy metals from acid mine drainage and other industrial effluents. In: Water Pollution and Remediation: Heavy Metals. Berlin/Heidelberg, Germany: Springer; 2021. pp. 513-537

[20] Abu Shmeis RM. Water chemistry and microbiology. Comprehensive Analytical Chemistry. 2018;**81**:1-56. DOI: 10.1016/bs.coac.2018.02.001

[21] Cyprowski M, Stobnicka-Kupiec A, Ławniczek-Wałczyk A, Bakal-Kijek A, Gołofit-Szymczak M, Górny RL. Anaerobic bacteria in wastewater treatment plant. International Archives of Occupational and Environmental Health. 2018;**91**(5):571-579. DOI: 10.1007/s00420-018-1307-6

[22] Roman MD, Sava C, Iluțiu-Varvara DA, Mare R, Pruteanu LL, Pică EM, et al. Biological activated sludge from wastewater treatment plant before and during the COVID-19 pandemic. International Journal of Environmental Research and Public Health. 2022;**19**(18):11323. DOI: 10.3390/ijerph191811323

[23] Orhon D. Evolution of the activated sludge process: The first 50 years. Journal of Chemical Technology & Biotechnology. 2014;**90**(4):608-640. DOI: 10.1002/jctb.4565

[24] Yang Y, Wang L, Xiang F, Zhao L, Qiao Z. Activated sludge microbial community and treatment performance of wastewater treatment plants in industrial and municipal zones. International Journal of Environmental Research and Public Health. 2020;**17**(2):436. DOI: 10.3390/ijerph17020436

[25] Ye C, Yang X, Zhao FJ, Ren L. The shift of the microbial community in activated sludge with calcium treatment and its implication to sludge settleability.

Bioresource Technology. 2016;**207**:11-18. DOI: 10.1016/j.biortech.2016.01.135

[26] Xia Y, Wen X, Zhang B, Yang Y. Diversity and assembly patterns of activated sludge microbial communities: A review. Biotechnology Advances. 2018;**36**:1038-1047. DOI: 10.1016/j.biotechadv.2018.03.005

[27] Fang H, Zhang H, Han L, Mei J, Ge Q, Long Z, et al. Exploring bacterial communities and biodegradation genes in activated sludge from pesticide wastewater treatment plants via metagenomic analysis. Environmental Pollution. 2018;**243**:1206-1216. DOI: 10.1016/j.envpol.2018.09.080

[28] Sharma M, Agarwal S, Agarwal Malik R, Kumar G, Pal DB, Mandal M, et al. Recent advances in microbial engineering approaches for wastewater treatment: A review. Bioengineered. 2023;**14**(1):2184518. DOI: 10.1080/21655979.2023.2184518

[29] Ajuna HB, Lim HI, Moon JH, Won SJ, Choub V, Choi SI, et al. The prospect of hydrolytic enzymes from bacillus species in the biological control of pests and diseases in forest and fruit tree production. International Journal of Molecular Sciences. 2023;**24**(23):16889. DOI: 10.3390/ijms242316889

[30] Shchegolkova NM, Krasnov GS, Belova AA, Dmitriev AA, Kharitonov SL, Klimina KM, et al. Microbial community structure of activated sludge in treatment plants with different wastewater compositions. Frontiers in Microbiology. 2016;**7**:90. DOI: 10.3389/fmicb.2016.00090

[31] Akram M. Mini-review on glycolysis and cancer. Journal of Cancer Education. 2013;**28**(3):454-457

[32] Granchi C, Bertini S, Macchia M, Minutolo F. Inhibitors of lactate dehydrogenase isoforms and their therapeutic potentials. Current Medicinal Chemistry. 2010;**17**(7):672-697

[33] Eren AM, Sogin ML, Morrison HG, Vineis JH, Fisher JC, Newton RJ, et al. A single genus in the gut microbiome reflects host preference and specificity. ISME Journal. 2015;**9**(1):90-100. DOI: 10.1038/ismej.2014.97

[34] Gao P, Xu W, Sontag P, Li X, Xue G, Liu T, et al. Correlating microbial community compositions with environmental factors in activated sludge from four full-scale municipal wastewater treatment plants in Shanghai, China. Applied Microbiology and Biotechnology. 2016;**100**(10):4663-4673. DOI: 10.1007/s00253-016-7307-0

[35] Ibarbalz FM, Orellana E, Figuerola ELM, Erijman L. Shotgun metagenomic profiles have a high capacity to discriminate samples of activated sludge according to wastewater type. Applied and Environmental Microbiology. 2016;**82**(15):5186-5196. DOI: 10.1128/AEM.00916-16

[36] Zeng J, Li J, Gou M, Xia ZY, Sun ZY, Tang YQ. Effective strategy for improving sludge treatment rate and microbial mechanisms during chromium bioleaching of tannery sludge. Process Biochemistry. 2019;**83**:159-167. DOI: 10.1016/j.procbio.2019.05.019

[37] Mohd Azhar SH, Abdulla R, Jambo SA, Marbawi H, Gansau JA, Mohd Faik AA, et al. Yeasts in sustainable bioethanol production: A review. Biochemistry and Biophysics Reports. 2017;**10**:52-61

[38] Liszkowska W, Berlowska J. Yeast fermentation at low temperatures: Adaptation to changing environmental conditions and formation of volatile compounds. Molecules. 2021;**26**(4):1035. DOI: 10.3390/molecules26041035

[39] Nzila A, Razzak SA, Zhu J. Bioaugmentation: An emerging strategy of industrial wastewater treatment for reuse and discharge. International Journal of Environmental Research and Public Health. 2016;**13**(9):846. DOI: 10.3390/ijerph13090846

[40] Fung KY, Lee CM, Ng KM, Wibowo C, Deng Z, Wei C. Process development of treatment plants for dyeing wastewater. AICHE Journal. 2012;**58**(10):2726-2742. DOI: 10.1002/aic.12798

[41] Ou D, Li H, Li W, Wu X, Wang YQ, Liu YD. Salt-tolerance aerobic granular sludge: Formation and microbial community characteristics. Bioresource Technology. 2018;**249**:132-138. DOI: 10.1016/j.biortech.2017.07.154

[42] Ali Shah F, Mahmood Q, Maroof Shah M, Pervez A, Ahmad Asad S. Microbial ecology of anaerobic digesters: The key players of anaerobiosis. The Scientific World Journal. 2014;**2014**:183752. DOI: 10.1155/2014/183752. Retraction in: The Scientific World Journal, 2017, 3852369

[43] Werkneh AA. Biogas impurities: Environmental and health implications, removal technologies and future perspectives. Heliyon. 2022;**8**(10):e10929. DOI: 10.1016/j.heliyon.2022.e10929

[44] Paglini R, Gandiglio M, Lanzini A. Technologies for deep biogas purification and use in zero-emission fuel cells systems – A review. Energies. 2022;**15**(3551):1-30

[45] Babaei A, Shayegan J. Effects of temperature and mixing modes on the performance of municipal solid waste anaerobic slurry digester. Journal of Environmental Health Science and Engineering. 2020;**17**(2):1077-1084. DOI: 10.1007/s40201-019-00422-6

[46] Chen XY, Vinh-Thang H, Ramirez AA, Rodrigue D, Kaliaguine S. Membrane gas separation technologies for biogas upgrading. RSC Advances. 2015;**5**:24399-24448

[47] Hoyer K, Hulteberg C, Svensson M, Jernberg J, Nørregård Ø. Biogas upgrading - technical review. Energiforsk. 2012;**40**:174-180

[48] Sudiartha GAW, Imai T, Hung YT. Effects of stepwise temperature shift in anaerobic digestion for treating municipal wastewater sludge: A genomic study. International Journal of Environmental Research and Public Health. 2022;**19**(9):5728. DOI: 10.3390/ijerph19095728

[49] Shabbir MA, Ahmed W, Khan MR, Ahmad T, Aadil RM. Valorization of agri-food wastes and by-products. In: Recent Trends, Innovations, and Sustainability Challenges. Cambridge, United States: Elsevier Science; 2021. pp. 645-663. DOI: 10.1016/B978-0-12-824044-1.00028-3

[50] Farghali M, Osman AI, Umetsu K, Rooney DW. Integration of biogas systems into a carbon zero and hydrogen economy: A review. Environmental Chemistry Letters. 2022;**20**:2853-2927. DOI: 10.1007/s10311-022-01468-z

[51] Zhu X, Zhou P, Chen Y, Liu X, Li D. The role of endogenous and exogenous hydrogen in the microbiology of biogas production systems. World Journal of Microbiology and Biotechnology. 2020;**36**:1-7

[52] Medina-Martos E, Istrate IR, Villamil JA, Gálvez-Martos JL, Dufour J, Mohedano ÁF. Techno-economic and life cycle assessment of an integrated

hydrothermal carbonization system for sewage sludge. Journal of Cleaner Production. 2020;**277**:122930

[53] Wilén BM, Liébana R, Persson F, Modin O, Hermansson M. The mechanisms of granulation of activated sludge in wastewater treatment, its optimization, and impact on effluent quality. Applied Microbiology and Biotechnology. 2018;**102**(12):5005-5020. DOI: 10.1007/s00253-018-8990-9

[54] Paolini V, Petracchini F, Segreto M, Tomassetti L, Naja N, Cecinato A. Environmental impact of biogas: A short review of current knowledge. Journal of Environmental Science and Health, Part A. 2018;**53**:899-906

[55] Mpongwana N, Rathilal S, Tetteh EK. Recovery strategies for heavy metal-inhibited biological nitrogen removal from wastewater treatment plants: A review. Microorganisms. 2022;**10**(9):1834. DOI: 10.3390/microorganisms10091834

[56] Alengebawy A, Abdelkhalek ST, Qureshi SR, Wang MQ. Heavy metals and pesticide toxicity in agricultural soil and plants: Ecological risks and human health implications. Toxics. 2021;**9**(3):42. DOI: 10.3390/toxics9030042

[57] Li J, Ding LB, Cai A, Huang GX, Horn H. Aerobic sludge granulation in a full-scale sequencing batch reactor. BioMed Research International. 2014;**2014**:268789. DOI: 10.1155/2014/268789

[58] Mahdizade Ari M, Dadgar L, Elahi Z, Ghanavati R, Taheri B. Genetically engineered microorganisms and their impact on human health. International Journal of Clinical Practice. 2024;**2024**:6638269. DOI: 10.1155/2024/6638269

[59] Coconi-Linares N, Magaña-Ortíz D, Guzmán-Ortiz DA, Fernández F, Loske AM, Gómez-Lim MA. High-yield production of manganese peroxidase, lignin peroxidase, and versatile peroxidase in Phanerochaete chrysosporium. Applied Microbiology and Biotechnology. 2014;**98**(22):9283-9294. DOI: 10.1007/s00253-014-6105-9

[60] Ramesh N, Ge Y, Ennist DL, Zhu M, Mina M, Ganesh S, et al. CG0070, a conditionally replicating granulocyte macrophage colony-stimulating factor-armed oncolytic adenovirus for the treatment of bladder cancer. Clinical Cancer Research. 2006;**12**(1):305-313. DOI: 10.1158/1078-0432.CCR-05-1059

[61] Parker JN, Pfister LA, Quenelle D, other authors. Genetically engineered herpes simplex viruses that express IL-12 or GM-CSF as vaccine candidates. Vaccine. 2006;**24**(10):1644-1652. DOI: 10.1016/j.vaccine.2005.09.051

[62] Aziz A, Basheer F, Sengar A, Irfanullah SUK, Farooqi IH. Biological wastewater treatment (anaerobic-aerobic) technologies for safe discharge of treated slaughterhouse and meat processing wastewater. Science of the Total Environment. 2019;**686**:681-708. DOI: 10.1016/j.scitotenv.2019.05.295

[63] Aziz A, Rameez H, Sengar A, Sharma D, Farooqi IH, Basheer F. Biogas production and nutrients removal from slaughterhouse wastewater using integrated anaerobic and aerobic granular intermittent SBRs - bioreactor stability and microbial dynamics. Science of the Total Environment. 2022;**848**:157575. DOI: 10.1016/j.scitotenv.2022.157575

[64] Khan NA, Singh S, Ramamurthy PC, Aljundi IH. Exploring nutrient removal mechanisms in column-type SBR with simultaneous nitrification and denitrification. Journal of Environmental Management. 2024;**349**:119485

Chapter 2

Microalgae and Black Soldier Fly Larvae as Sustainable Methods for Decentralized Sewage Treatment in Sub-Saharan Africa

Ali Shee, Mariam Maku Swaleh, Laila Uweso Abubakar, Huxley Mae Makonde, Saeed Mutta Mwaguni, Daniel Munga, Mumini Ndenge Dzoga and Cyrilus Otulo Wandera

Abstract

Human population in many African countries is on upward trajectory resulting in increased waste generation. Currently, the generation of human wastes supersedes their collection, treatment and disposal. Sewage management is therefore a major problem. Available traditional sewage management systems comprise of pit latrines, soak pits, cesspools, and septic tank-soakage pits. Non-traditional, but conventional systems include constructed wetlands (CWs) and municipal wastewater treatment plants (MWWTPs). Whereas the former are largely ineffective, CWs and MWWTPs achieve significant detoxification enabling reuse of sludge and effluent water, but require considerable land sizes coupled with high investments in capital, operational and maintenance costs. Hence are less common. Pit-latrines, soak pits, cesspools and septic tank-soakage pits are locally preferred due to their lower construction and repair costs, but ineffective in sewage treatment. Herein, we present the opportunities of using black soldier fly larvae (BSFL) and microalgae as cost-effective and sustainable sewage treatment methods. A deeper understanding on the technicalities and suitability of BSFL and microalgae is provided. Microalgae are tailored for secondary wastewater treatment into high-quality effluent for reuse or discharge into the environment. Accumulated microalgae biomass is convertible into bio-oil, biofertiliser and biofuel. BSFL is relevant for primary sewage sludge treatment producing nutrient-rich frass as biofertilizer. The pupae are rich in protein, fats and fatty acids hence usable as animal feed. Significant gains are obtained by combining BSFL and microalgae in batch processes employing pump and treat. The process requirements, environmental and socio-economic benefits of these methods are presented to guide in decision making.

Keywords: black soldier fly larvae (BSFL), decentralized sewage treatment, microalgae, sewage assessment, sewage sludge, wastewater treatment

1. Introduction

Many countries in Africa are experiencing sustained economic growth rates and urbanization. Across the globe, Africa has a young population with a median age < 24 years [1, 2]. It is projected that by 2050, Africa will be home to nearly 30% of the global population [2]. Currently, Africa's population is spread across rural villages, peri-urban areas, municipalities, and cities [1, 3]. Globally, Africa has the least number of urban dwellers. However, by 2050, more than 60% of Africa's population will be living in urban areas [4]. The rapid increase in population and urbanization coupled with decreased mortality rates presents many challenges that can be harnessed into opportunities. Proper sewage management and adequate sanitation services are a challenge common across many countries in Africa especially sub-Saharan Africa [5]. For example, in Eastern and Southern Africa, approximately 340 million people or 70% of the population lack access to basic sanitation facilities [5, 6]. This population segment practices open defecation, and use unimproved or shared facilities. Sanitation facilities and personal hygiene are critical in ensuring a healthy society. Hence, they must not be looked at in isolation but wholesome. The lack of any of these systems has implications for public health. Their inadequacy has contributed to the sporadic spread of waterborne diseases such as cholera, typhoid, dysentery, and bilharzia. Frequent sickness due to some of these diseases is associated with increased child mortality, stunted growth in children, poor performance in school, and diminished productivity by employees [5, 6].

Sewerage systems have multiple purposes in developing countries. Urban areas and cities have sewerage systems that serve as drainage systems as well. In addition to this, wastewater from domestic, industries, and other sources are channeled through the same system. This broadens the understanding of the sewage concept in developing countries to include wastewater from different sectors transported and disposed of using an established system. Therefore, sewage may be defined as effluent emanating from human excretes, domestic, industries, hospitals, factories, and storm runoff waters being collected, treated, and disposed of altogether. In sub-Saharan Africa, common sewerage systems comprise pit latrines, soak pits, cesspools, septic tanks, constructed wetlands (CWs), and municipal wastewater treatment plants (MWWTPs) [7–11]. Pit latrines, soak pits, and cesspools are cheap, and easy to install and repair hence are widely used by rural and peri-urban populations [5, 6]. Septic tank-soakage pit systems are slightly expensive to install, hence mostly found in high-end urban centers [12, 13]. CWs are cheap to operate but require huge tracts of land and high initial capital [14]. Similarly, MWWTPs require huge land masses coupled with elaborate pipe networks and drainage systems containing suitable treatment chambers. Due to their high investment costs, government support for MWWTPs is minimal, hence they are few, and where found, they are dilapidated and ineffective [15, 16]. Poor maintenance of MWWTPs results to seepages and over-flows particularly during rainy seasons. It is observed that most of sub-Saharan African countries have open sewers channeled through surface water sources such as rivers and oceans with severe ecological implications [17, 18].

Sewage consists mostly of wastewater (>95%), solids (<1%) and other dissolved or suspended substances that may include bacteria, viruses, pathogens, nutrient pollutants (nitrates, nitrites, ammonia phosphate, etc.), heavy metals, and emerging water contaminants (EWCs) [5, 6, 10, 19]. EWCs are substances that are present in wastewater but have not been of interest to monitoring by water treatment experts and environmental agencies [20]. Some EWCs include pesticides, antibiotics, cosmetic products, pharmaceuticals, and veterinary drugs. Due to their physical, chemical, and bioaccumulation nature, EWCs are potentially toxic and may be carcinogenic. Many EWCs are also resistant to biodegradation and thus persist in the environment [20].

In sub-Saharan Africa, less than 1% of sewage is treated [5, 7, 16]. Thus, raw sewage is directly released into rivers, lakes, oceans, and the environment [7, 17, 21]. The presence of various chemical substances in sewage is responsible for the dark appearance, unpleasant odors, eutrophication, and environmental pollution. This puts both the public and ecological health at risk. Also, raw sewage is a source of greenhouse gases such as CH_4 and CO_2 [22, 23]. In order to protect public health, sewage treatment and detoxification is important. To ensure sustainable resource utilization, recycling the effluent sludge and water is vital in order to recoup some of the treatment costs. Presently, the recovery of valuable nutrients such as nitrogen from wastewater is essential. Several factors are critical during the selection of any treatment system. These include the economic ability of the user, availability of government and donor support, population size, geographical location, ease of operation and maintenance, detoxification efficiency, and the possibility recovering treated water, energy, and nutrients. This chapter highlights the common sewerage treatment systems available in sub-Saharan Africa, their advantages, disadvantages, and site selection suitability. Also presented is a deeper insight and the potential of black soldier fly larvae (BSFL) and microalgae in sewage management and treatment.

2. Common sewerage systems in sub-Saharan Africa

As discussed earlier, the common sewerage systems in low- and middle-income economies comprise pit latrines, soak pits, cesspools, septic tanks, CWs, and MWWTPs [5, 7, 14, 16, 24]. **Table 1** below describes the key features of these systems, the requirements for installation, and their advantages and disadvantages.

As indicated in **Table 1**, CWs and MWWTPs require huge land masses and capital investments. Hence, they are less common in many low- and middle-income countries like Kenya. Pit latrines, soak pits, cesspools, and septic tanks are less efficient in reducing organic load and removal of pollutants. Hence, further treatment of the sludge, solids, and wastewater is necessary. Effective waste management strategies are essential to ensure that the increase in waste generated does not overwhelm local infrastructure. Waste management encompasses the collection, treatment, and disposal of waste materials, and it is particularly challenging in informal settings due to the high volumes and varied types of waste produced. Also, informal settings are poorly designed lacking critical infrastructures such as roads, drainage, electricity, and freshwater supply. In order to effectively manage sewage systems, the critical criteria to be considered include assessment of discharge volume, composition, desired effluent quality, efficiency of the treatment system, and resources sustainability.

3. Sewage assessment

Sewage assessment encompasses the systematic measurement and documentation of wastewater generation and treatment processes [34, 35]. It involves tracking the amount of sewage production, the efficiency of treatment systems, and the environmental effects and impacts of the sewage disposed. Obtaining this data is essential for assessing the effectiveness of waste management systems and strategies and, in

System type	Description	Operation	Suitability and site selection	Advantages	Disadvantages	Reference
Pit latrine	A large hole in the ground covered with a slab having an opening where the user excretes.	Urine, water, and fecal fall into the pit. If the walls are porous, urine and water sips into the ground while fecal matter and other solids remain. The fresh fecal matter is decomposed by aerobic bacteria and then by anaerobic bacteria.	Domestic households or small development set-ups. The ground should have sufficient porosity for liquids to soak away.	Easily built and repaired using locally available materials. Requires less capital investments and operating costs. Requires small land area.	Bad odors. Clogging can lead to overflow. Sewage undergoes very minimal treatment. Unsuitable for large developments. Risks of pollution of underground aquifers.	[25, 26]
Soak pit	A covered chamber lined with porous walls and porous base allows water to slowly soak into the ground.	A pre-settled effluent from a holding/treatment area is channeled to the underground chamber allowing liquids to infiltrate into the ground.	Single family settings. Small-scale development set-ups. The ground site should have sufficient porosity for liquids to soak away.	Low costs for installation and repair. Easily built with locally available materials. Requires small land area.	Need to undertake primary treatment to minimize clogging. Unsuitable for areas experiencing heavy rainfall and floods. Need for separation of greywater and black water. Minimal sludge treatment occurs. Inappropriate to recycle wastewater.	[27–29]
Cesspool	A holding tank that is placed to hold sewage for short periods of time.	Raw sewage flows into the tank and accumulates before being siphoned and transported away.	Single family households. Small-scale development. Suitable for sites where the water table is close to the surface. Suitable where impermeable rocks or clay soils hinder liquids infiltrating into the ground.	Minimal costs for installation, maintenance, and repair.	High risk of pollution when the tank is damaged or not properly handled. Risk of overflow and contamination. Unsuitable for large-scale developments. Bad odors. Scheduled exhauster costs.	[27–29]

System type	Description	Operation	Suitability and site selection	Advantages	Disadvantages	Reference
Septic tank	A watertight tank or chamber designed to hold sewage solids at the bottom while draining the wastewater through an outlet.	Raw sewage flows into the tank where solids settle at the bottom and liquids float and flow out through an outlet by gravity to a nearby land where it soaks away. Some of the sludge is degraded by anaerobic bacteria.	Single use domestic households. Industrial set ups. Small-scale development. The site should allow infiltration of effluents into the ground.	Ease to install Some capital is required for installation, maintenance, and repair. Minimal sewage treatment occurs. No need for electricity compared to MWWTPs.	Minimal sewage treatment increases the risks of pollution. Wastewater contamination of nearby surfaces. Risk of flooding, overflow, and backflow in cases of poor site porosity. Bad odors. Scheduled exhauster costs.	[30, 31]
Constructed wetland	A facility that applies the natural functions of wetland vegetation, soil, sediments, and microbial assemblies to treat sewage.	Raw sewage is passed through artificially constructed shallow ponds where wetland vegetation, microorganisms and sometimes aquatic animals are introduced to treat sewage.	Suitable for large-scale development. The site should allow easy flow of wastewater.	Requires minimal energy input. Low operational and maintenance costs. Removes most contaminants, pollutants, and organic load enabling the reuse of the effluent water. Provide habitat wetlands vegetation and aquatic organisms. Easily constructed according to landscape. Requires no electricity.	Requires huge tracks of land for installation. High initial investment costs needed. Hindered by high concentration of pollutants. May act as breeding sites for insects, pathogens, and vectors.	[32, 33]
MWWTPs	Facilitates that enable the preliminary, primary, secondary, tertiary, and advanced treatment of sewage into sludge and wastewater that meet environmental quality standards.	Raw sewage is passed through a combination of physical, chemical, electrochemical, and biological processes. These processes remove most pollutants, organic load, microorganisms, and contaminants. The effluent water can be recycled while the sludge may be converted into high-value products.	Suitable for large developments. Location must be easily accessible. Site should easily allow the installation of pipe networks and drainage lines.	Achieves significant treatment of sewage. Effluent water can be reused. Treated sludge can be reused as fertilizer or production of high-value products.	Huge capital investments needed. Require skilled personnel for operation, maintenance, and repair. Requires electricity for aeration. Micropollutants such as EWCs may escape untreated.	

Table 1.
Sewerage management and treatment systems and their site suitability, advantages, and disadvantages.

planning for improvements. Analyzing data on sewage generation and treatment supports trend identification, needs forecasting, and well-informed infrastructure upgrade decision-making. The amount of sewage generated can be measured using various tools such as flow meters or flow sensors. Most of these are designed for regions with elaborate sewer lines to indicate the amount of effluent moving through a pipe or conduit by measuring linear, nonlinear, mass, or volumetric flow rates. In establishments using pit latrines, cesspools, CWs, soak pits, and septic tank-soakage pit systems, measuring volumes of sewage is challenging. These systems hold sewage, allowing in some of them deposition and separation of wastewater. For such systems if properly designed flow meters and sensors can be used. Also, for cesspools and septic tanks, quantification can be done when sewage is siphoned by exhausters.

The performance of the treatment system is determined by calculating removal efficiency (RE: the amount of contaminant removed through the treatment process). A contaminant in this case is any parameter above the required limits. RE can be calculated for a single unit or for an entire system comprising several treatment units. Before discharge, the components in the effluent are determined and its limits adjusted to meet environmental quality standards. Each country usually has their own effluent quality limits mainly adopted by the United States Environmental Protection Agency [36] and the World Health Organization [24]. Based on the operating environment and the nature of the contaminant, a single or a combination of processes are applied to achieve desired effluent quality limits.

4. Alternative sewage treatment systems applicable to sub-Saharan Africa

4.1 Black soldier fly larvae

Black soldier fly (BSF) or *Hermetia illucens* is an insect belonging to the Diptera family and Stratiomyidae order. It thrives well in the tropics and warmer temperate regions. The mature flies are harmless to humans and are not attracted to food. They are also noninvasive and non-biting. Hence, they are not a threat to humans and the environment. As shown in **Figure 1**, BSF has four growth stages lasting about 45 days (egg-4 days, larvae-18 days, pupae-14 days, and adult-9 days) [37, 38].

Black soldier fly larvae (BSFL) are the most important in organic waste management. BSFL are able to reduce huge volumes of biological and organic wastes into nutrient-rich frass and high-end protein. Several studies have underpinned the significance of BSFL in recycling of various types of biological and organic wastes such as kitchen waste, pig manure, food waste, abattoir waste, fruit waste, and vegetable waste [39–42].

4.1.1 Operating parameters for BSFL

Temperature is critical in BSFL performance and it should range between 25°C and 30°C [42, 43]. Other process parameters include humidity, oxygen content, and nutrient composition. BSFL growth requires an adequate supply of nutrients; hence the organic substrate selected must contain higher amounts of digestive nutrients, proteins, and carbohydrates [42]. Nutrient-deficient substrates hinder BSFL growth and performance [44, 45]. Under optimal temperature conditions, BSFL can attain >75% bioconversion performance of raw organic waste into nutrient-rich frass. The pupae are rich in protein, fats, and fatty acids, hence suitable for use as animal feed.

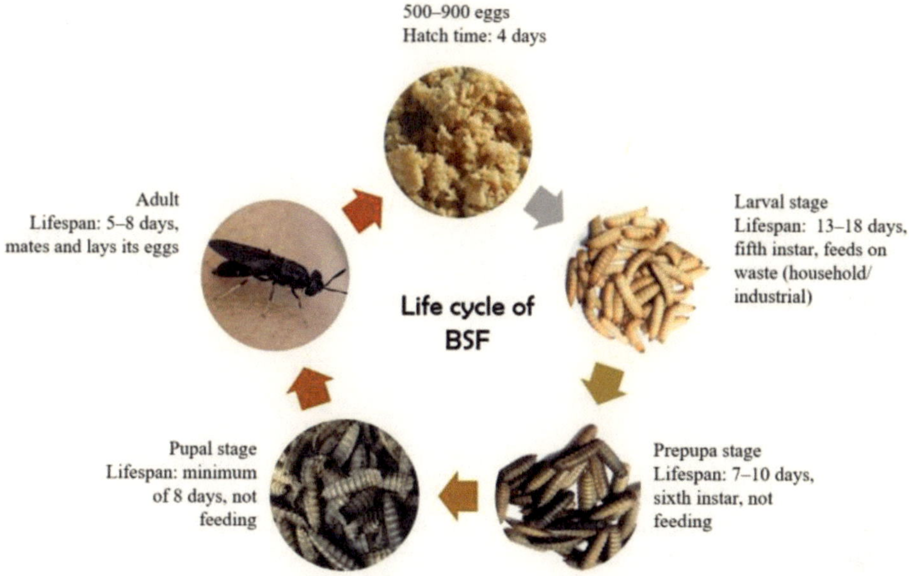

Figure 1.
Life cycle of black soldier fly [37].

4.1.2 Application of BSFL in sewage sludge treatment

The potential of BSFL is also seen in sewage treatment [46]. BSFL can be applied both in fecal sludge digestion as well as the removal of pollutants in wastewater [39]. In the treatment of fecal sludge using BSFL, dewatering may not be necessary [46, 47]. Hence, BSFL can be used directly for primary sewage treatment. The evaluation criteria to determine the fecal sludge treatment efficiency using BSFL involves calculating fecal sludge reduction (FSR) rate (%). FSR is calculated according to Eq. (1) [48].

$$FSR = \frac{\text{Initial weight} - \text{Final weight}}{\text{Initial weight}} \times 100 \qquad (1)$$

Tokwaro et al. [48], obtained an FSR of 66% and 72% for unlined and lined pit latrines, respectively. Similarly, Banks [47] obtained an FSR of 50% and 55% using pit latrine sludge and fresh feces, respectively. Lalander et al. [46] also reported an FSR of 73% while Achieng Oyoo et al. [44] reported FSR between 76% and 92%. These rates are higher than previous studies due to the introduction of kitchen waste in the feed. The kitchen waste provides nutrients to the larvae hence improved performance. Sewage sludge pre-treatment using anaerobic bacteria permits the rapture of extracellular polymeric substances (EPS) thereby releasing entrapped nutrients. Even in the absence of kitchen wastes and pre-treatment, waste reduction rates by BSFL for sewage sludge are similar to waste reduction rates for common types of biological and organic wastes [44, 49, 50]. These studies highlight the potential role of BSFL for *in situ* treatment of fecal sludge. The advantage of using BSFL in fecal sludge is also seen in the removal of harmful microorganisms such as Salmonella spp. [46]. Nevertheless, BSFL fail to remove helminth eggs (Ascaris ova), hence need for further treatment.

4.1.3 BSFL application in wastewater treatment

BSFL has been utilized in the removal of water and environmental contaminants such as heavy metals, polycyclic aromatic hydrocarbons (PAHs), and antibiotics. Heavy metals, PAHs and EWCs such as antibiotics are difficult to treat by conventional techniques. Since BSFL is able to sequester these substances from sewage, the effluent is able to meet environmental standards for discharge and reuse [49, 51]. Also, it is noteworthy to point out that these contaminants once ingested do not exceed feed limits. Hence, upon sequestration of these water contaminants, BSF pupae can still be used in the preparation of animal feeds and other products of economic value [49, 51].

4.1.4 BSFL's potential application for sewage treatment in sub-Saharan Africa

Sewage contains about 95% wastewater, <1% fecal sludge with the remaining comprising solids and other substances. Therefore, the treatment and recycling of wastewater is essential. Among the available sewage systems in sub-Saharan African countries, CWs and MWWTPs can achieve significant wastewater detoxification and are available for reuse and irrigation. These facilities are expensive to set up, run, and maintain hence rare. In small-scale developments, wastewater could be pumped out from cesspools and septic tanks for treatment and reuse. However, it will be challenging for pit latrines and soak pits. The presence of permeable barriers causes wastewater to infiltrate into the ground. Treatment of wastewater collected from cesspools and septic tanks using conventional processes such as membrane filtration is expensive for small scale.

A cheaper and more sustainable option for treatment of sewage in pit latrines, soak pits, cesspools, and septic tanks is presented by BSFL. The practicality can be seen using pump and treat technologies. The approach will involve regular pumping of sewage to an above-ground reactor where fecal sludge and wastewater are separated. This separation is essential to increase the concentration of the sludge for degradation by BSFL. In the presence of wastewater, fecal sludge, and other solids will sediment at the bottom of the reactor. In the case of an agitated reactor, the substrates will be dispersed in wastewater. |Thus, direct contact between substrates and BSFL will be minimal decreasing treatment efficiency. BSFL is then introduced into the reactor containing concentrated fecal sludge. The reactor should be kept under controlled conditions of temperature (25–30°C) to support the growth of BSFL. Air should be applied to promote larval growth. In the absence of air, larval mortality occurs. Also, the reactor must be fitted with a suitable stirrer to ensure uniform mixing of BSFL and the substrates. In our opinion, BSFL can as well be introduced into shallow pit latrines and soak pits that are well-aerated. However, it will be difficult for the recovery of products. Since nutrients are usually entrapped in EPS, the mixing of sewage with kitchen, fruit, and vegetable wastes is essential to obtain high FSR rates. The organic substrates supply essential nutrients for BSFL growth. Therefore, BSFL can be used for the simultaneous treatment of fecal sludge, wastewater, and other wastes from fruits, vegetables, and kitchen. The organic matter upon conversion into frass can be applied as biofertilizer. The effluent water can be recycled or used for irrigation. On the other hand, the pupae can be used as animal feed and production of other valuable substances such as biofuels.

The major drawback of the pump and treat system using BSFL is the pumping cost. For small-scale developments, pumping costs can be shared. Also, consideration for solar pumps especially in the tropics is attractive. The application of BSFL for decentralized sewage treatment requires minimal land and therefore pose no major threat to the use of land for other economic activities such as farming. In summary, BSFL provides a robust, sustainable, and environmentally compatible technique for simultaneous primary treatment of sewage and other types of biological and organic wastes.

4.2 Microalgae

Microalgae are microscopic photosynthetic plants from a heterogeneous group called the algae. They live in almost every habitat; however, they are mostly found in aquatic ecosystems either in fresh or salty water. They can also be found in deserts, volcanic waters, and highly acidic and frozen soils [52, 53]. Microalgae can either be prokaryotic or eukaryotic, which means they can exist as single-celled entities with a simple cell structure or have a simple multicellular arrangement [52, 54]. However, most microalgae are eukaryotic, with the exception of a few prokaryotic species such as cyanobacteria (blue-green algae). Microalgae are very small in size and a microscope is required for visualization [52, 54, 55]. The microalgae cells range in size from 2 to 200 micrometers and can exist in various forms, including polyphyletic and noncohesive types. The diverse structures and sizes of microalgae enable them to adapt to a wide range of environments and proliferate rapidly [52]. Microalgae can survive with minimal water, nutrients, and carbon dioxide while efficiently utilizing solar energy through photosynthesis [54]. They are highly diverse with more than 30,000 species documented [52].

Bacillariophyta (Diatoms) are the most abundant group of microalgae in terms of biomass and distribution. They are found in a wide range of aquatic environments. Diatoms are a dominant group of phytoplanktons in both marine and freshwater environments, contributing significantly to primary production in aquatic ecosystems [56, 57]. They are responsible for a substantial portion of global photosynthesis and are key players in the carbon cycle due to their high biomass and widespread distribution [56].

Microalgae are a critical component of aquatic ecosystems, serving as the foundation of the food chain. Positioned at the base, they are consumed by zooplankton, which in turn are preyed upon by fish [55]. As primary producers, microalgae contribute approximately 40% of global photosynthesis [53]. Their biomass and composition significantly influence primary productivity in aquatic environments. Consequently, changes in microalgae biodiversity and abundance can disrupt the aquatic food web and impact overall biological productivity [57, 58]. Microalgae productivity is influenced by environmental factors such as availability of nutrients (phosphorus and nitrogen), temperature, salinity, humidity, pH, and light intensity [59–61]. Despite these variables, many microalgae species are adaptable and can thrive in a range of conditions due to their resilience to fluctuations in these environmental parameters [54]. This adaptability is crucial for their application in wastewater treatment, where they are utilized to address various environmental challenges.

4.2.1 Utilization of microalgae for wastewater treatment

The use of microalgae in wastewater treatment (WWT) dates back to the 1900s dates back to the 1900s, where scientist observed and recognized the potential of microalgae in improving water quality. Ever since, microalgae have been cultivated in various wastewater mediums such as agricultural, industrial, domestic, and municipal.

Researchers established that microalgae play a crucial role in wastewater treatment by absorbing both organic and inorganic nutrients from the wastewater. This is because they thrive well in ecosystems with large amounts of nitrogen and phosphorus, as they use these nutrients for their growth [57]. It has been shown that they can remove nitrogen and phosphorus from wastewater to efficiencies of up to 80–100%. Laboratory and field studies have also shown that microalgae are cost-effective in the removal of biological oxygen demand (BOD), chemical oxygen demand (COD), and pathogens. Microalgae can also assimilate heavy metals and organic pollutants like pharmaceutical products from wastewater [62].

Wastewater treatment efforts have predominantly focused on freshwater microalgae, even though many coastal towns struggle with wastewater management, often leading to untreated or inadequately treated wastewater being released into surface waters such as lakes, rivers, and the ocean. This practice results in problems such as eutrophication and the creation of dead zones [17]. Marine microalgae present a promising alternative. Research shows that marine microalgae wastewater treatment systems can reduce freshwater needs by up to 90% and are capable of withstanding harsh environmental conditions and wastewater of high salinity levels such as agricultural runoff [63]. Marine microalgae, such as *Chlorella*, *Spirulina*, and *Nannochloropsis*, are frequently used for wastewater treatment due to their high nutrient uptake efficiency and adaptability to various environments [64, 65]. Marine *Chlorella*, *Oscillatoria*, and *Entomoneis* species, for instance, are known for their ability to remove nitrogen and phosphorus from wastewater [57]. A study by Craggs et al. [64] demonstrated that the marine microalgae *Phaeodactylum tricornutum* and *Oscillatoria* sp. were capable of removing up to 100% ammonium and orthophosphate from wastewater.

Microalgae-based wastewater treatment presents numerous advantages, making it an appealing option for developing countries with decentralized sewage systems. Compared to conventional methods, microalgae treatment offers several benefits, such as reduced energy consumption, minimal chemical usage, and the production of valuable biomass [66]. Production of large amounts of sludge has been a major concern in other wastewater treatment systems. In conventional systems, hazardous chemicals are used to treat wastewater either in the correction of pH or the removal of odor and color. These chemicals result in huge amounts of potentially dangerous sludge that is eventually disposed of in landfills. The sludge in microalgae wastewater treatment contains useful biomass that can be processed to make biofuel or other valuable products. In addition, microalgae cultivation for wastewater treatment requires low maintenance, and minimum land use and is capable of converting CO_2 during photosynthesis to useful biomass for production of biofuels, animal feeds, etc. [55]. Additionally, they can be integrated with other treatment processes, such as algae-based biofilters and constructed wetlands [67].

Microalgae-based wastewater treatment systems are crucial for mitigating climate change due to their ability to sequester atmospheric CO_2 through photosynthesis [62]. Unlike terrestrial plants, microalgae exhibit significantly higher carbon fixation capacity, ranging from 10 to 50 times. Furthermore, the entirety of microalgae biomass can be harvested, offering a complete utilization of the captured carbon and recycling the treated water [68]. One of the most advantageous aspects of microalgae is their versatility in growth environments. They can thrive in various media, including saline water, and do not compete with agricultural crops for land resources. This characteristic allows microalgae to be cultivated in areas unsuitable for traditional farming, minimizing land use, and reducing competition for arable land [69].

However, the application of microalgae in wastewater treatment is primarily confined to tertiary treatment stages, where wastewater has already undergone significant pre-treatment. Raw sewage has high organic matter content and complex composition, which can inhibit microalgae growth and reduce their treatment efficiency. Microalgae generally perform better in treated or partially treated wastewater where organic loads are lower. Furthermore, maintaining optimal conditions for microalgae growth, such as light intensity and temperature, can be challenging in large-scale operations. Variations in these conditions can affect the treatment efficiency.

4.2.2 Potential of microalgae in sewage wastewater treatment

To enhance the efficiency of microalgae in wastewater treatment, several strategies have been employed.

4.2.3 Culturing of microalgae

Microalgae can be cultured in various settings, including open ponds, photobioreactors, and enclosed systems. Culturing conditions such as light intensity, temperature, and nutrient availability are crucial for optimizing growth and biomass yield. The cultivation of microalgae at a laboratory scale is performed using photobioreactors. Photobioreactors are bioreactor systems used in culturing microalgae in an enclosed system. These systems effectively prevent contamination and regulate environmental parameters, making them ideal for cultivating pure strains of microalgae and monocultures, thus ensuring high biomass yields. Additionally, photobioreactors are compact and require less land. However, they are not economically feasible for developing countries due to their high installation and maintenance costs. Conversely, open ponds are the most commonly used because they are more cost-effective and energy-efficient [70]. Open ponds are cheaper to construct, maintain, and operate. They come in various forms, including natural water bodies like lakes and ponds, and artificial ones such as circular and raceway ponds. In some cases, tanks can also be used to culture microalgae [71].

Open ponds containing microalgae can be considered more appropriate for decentralized treatment of sewage. Microalgae require oxygen for respiration and therefore may be ineffective in anaerobic environments. If anaerobic conditions exist, then bacteria-algal consortia should be considered. The synergy involves the production of oxygen by anaerobic bacteria which is consumed by microalgae for growth [72]. However, microalgae are also autotrophic hence their growth rates will be inhibited under light-deficient conditions. Since high concentration of organic load affects treatment efficiency by microalgae, pre-treatment techniques must be employed. The treatment approach will involve the separation of the fecal sludge and wastewater. The microalgae can be used for secondary and tertiary wastewater treatment to remove organic load, COD, BOD, organic micropollutants, heavy metals, and nutrients such as phosphorus and nitrogen.

4.2.4 Mono culturing and mixed cultures

A study by Hajri et al. [73] utilized mixed cultures of microalgae-*Spirulina platensis*, *Micractinium*, and *Chlorella*-for nutrient removal from dairy wastewater (DW). Microalgae were isolated and cultivated under various light conditions. The findings indicated significant biomass production, with mixed cultures achieving the highest biomass (2.51 g/L), followed by *Spirulina* (1.98 g/L) and *Chlorella* (1.92 g/L). Supplementing DW (75%) with blue-green medium (25%) significantly enhanced biomass and pH levels, improving pathogenic bacteria removal [73]. *Spirulina* and mixed cultures exhibited high nitrogen removal efficiencies of 92.56% and 93.34%, respectively, while *Chlorella* achieved 86.85% nitrogen and 83.45% phosphorus removal. Microbial and coliform removal efficiencies reached 97.81%, with elevated pH levels contributing to significant reductions in faecal *Escherichia coli* and coliform levels. The results demonstrate that integrating microalgae cultivation into treatment systems can significantly enhance nutrient and pathogen removal, providing a sustainable solution for wastewater management.

4.2.5 Co-culturing of microalgae for wastewater treatment

The combination of microalgae with activated sludge, fungi, bacteria, and nanoparticles benefits from a co-culture system during bioremediation. The possible co-culture systems include:

4.2.5.1 Bacteria-microalgae consortium

These include cultivating mixed cultures of microalgae rather than pure strains and integrating bacteria to support pollutant removal. Interactions between microalgae and bacteria can range from symbiotic to competitive. In symbiotic relationships, these interactions involve nutrient exchange, cell-to-cell communication, and the simulation of chemical compounds [74]. Additionally, a key aspect of the symbiotic relationship involves the exchange of gases. Microalgae produce oxygen through photosynthesis, which benefits heterotrophic bacteria by aiding in the oxidation of organic carbon. In turn, these bacteria release CO_2, which microalgae utilize for their photosynthetic processes [72]. Moreover, these systems require low energy compared to conventional systems where 45–75% of treatment costs are associated with energy used for mechanical aeration to provide oxygen to aerobic bacteria [75]. It is estimated that every kilogram (kg) of BOD removed consumes 1 kilowatt hour (kWh) of electricity for aeration [75]. However, no energy input is required for the removal of BOD by microalgae. The resultant microalgae biomass for removal of a kg of BOD is sufficient for the production of one kWh of electricity [75]. The interaction between microalgae and bacteria can stabilize the wastewater treatment process, making it more resilient to variations in wastewater composition and organic load [72]. Microalgae and bacteria consortiums are more effective than traditional methods of wastewater treatment. The system reduces CO_2 emissions and needs low energy inputs [76]. Pilot-scale and full-scale systems have demonstrated that these consortia can handle varying organic loads and contribute to more stable and efficient treatment processes [77, 78].

The bacteria and microalgae consortium can be put into a biogranule system composed of aggregated cells of microalgae and bacteria. This system provides enhanced high-treatment performance. The granules are designed so that they have a high settling velocity that is important in the regulation of biosolids/liquid separation and effluent quality from MWWTPs. The structure of granules also enhances the degradation and removal of contaminants, because the structure determines how long the microorganisms can interact with pollutants [79]. The algal-bacterial granules when used in wastewater treatment, have been reported to have low energy demands and simultaneous removal of carbon, nitrogen, and phosphorus, which was further discovered that the amount of greenhouse gas emissions is lower compared to an activated sludge system [69]. In addition, wastewater is considered a resource, especially in areas where resources are scare, the algal-bacterial system of WWT has high energy and material recovery including biogas, volatile fatty acids, biodiesel, bioplastics, and biopolymers [80]. However, most studies have been conducted only at a laboratory-scale level in controlled environments due to technological challenges that limit upscaling to pilot plants [80, 81].

4.2.5.2 Microalgae-fungi

Many studies have also investigated co-culturing microalgae with fungi and their symbiotic role in wastewater bioremediation. The synergic interaction between microalgae and fungi/yeast is observed as generating O_2 by microalgae for biomass synthesis. Then yeast utilizes O_2 for cellular respiration resulting in carbon substance disintegration releasing CO_2 required for microalgal photosynthesis. Walls et al. [82] co-cultured microalgae with yeast supplemented with glucose to achieve high biomass production and nutrient removal rates. The heterotrophic cultivation with 10 and 20 g L glucose yielded 1.85 and 2.74 g/L biomass concentration and removed 91% and 94% orthophosphate, 93% and 97% nitrate, and 93% and 95% total NH_4^+-N in 3 days, respectively [82]. Literature indicates that co-culturing *Chlorella* sp. with *Penicillium* sp. by treating the byproduct of the hydrothermal carbonization of wet biomass can achieve removal efficiency of 46.13% COD, 13% total nitrogen (TN), 6% NH_4^+-N, and 88.4% total phosphorus (TP) [83]. Therefore, the co-cultivation system performs better relative to a mono-system of microalgae in removing nutrients in wastewater.

4.2.5.3 Use of microalgae-activated sludge

The microalgal assimilation of nutrients and COD removal enhancement via activated sludge, signify the superiority of microalgae-activated sludge association in WWT, compared to the conventional mono-systems [84]. The wastewater characteristics and composition dictate the sludge/microalgae ratio. In a study by Mujtaba et al. [85], the co-culture of suspended activated sludge and immobilized *Chlorella vulgaris* in a single reactor led to an enhancement in nutrients removal with the decrease of inoculum ratio of activated sludge/*C. vulgaris* reaching an utmost value of 0.5% with nitrogen and phosphorus removals of 99.8% and 100%, respectively, within 2 days. Again, in municipal wastewater, at a low sludge/microalgae ratio (0.5), the growth of microalgae doubled, and more nutrient removal was achieved (66% removal of N and 100% of P in 1 day) relative to pure microalgal culture [86]. The microalgae/activated sludge ratio of 3:1 was selected to attain efficient organics and nutrient removal (86% TN, 70% TP, and 99% COD) in a mixed culture of a photobioreactor [84].

4.2.5.4 Use of microalgae-nanoparticles

Nanoparticles have numerous advantages of high surface-to-volume ratio, porosity, and biocompatibility. Thus, nanofibers are regarded as ideal supporting carriers over other conventional supporting materials for the immobilization and encapsulation of microbes, including microalgae, to produce bio-integrated hybrid materials with higher efficiency of pollutants removal, ease of application, and reusability [87]. A cross-linked chitosan nanofiber mat was used in immobilizing microalga *C. vulgaris* as a nontoxic support surface exhibiting diverse advantages associated with NO_3^- removal from water and combined with microalgal harvesting, dewatering, and processing steps into a single stage [88]. The microalgal-immobilized chitosan nanofiber featured simplicity, robustness with 6 months of contact time with water, and a higher elimination rate of nitrate (87 ± 4%) compared with 32 ± 3% for non-immobilized nanofiber. Vasistha et al. [89] utilized microalgae with ZnO nanoparticles to evaluate nutrient reductions from primary and secondary treated sewage WW. The co-cultured bionanofiber achieved removal efficiencies of 97.5%, 87.20%, 82.21% and 95.3%, 85.2%, 81.5% for total organic carbon (TOC), TN, and TP from primary and secondary treated sewage WW, respectively [89].

4.2.6 Proposed sewage treatment system using BSFL and microalgae in sub-Saharan Africa

As discussed earlier, the common sewerage systems in low- and middle-income economies involve pit latrines, soak pits, cesspools, and septic tanks. Pit latrines and soak pits designs are confined with poor illumination and minimal air circulation. Furthermore, most of the wastewater infiltrates through permeable surfaces into the ground. Thus, it will be difficult to use microalgae in such cases. For the treatment of sewage from cesspools and septic tanks using microalgae, pump and treat techniques are ideal. In this case, sewage from cesspools and septic tanks is pumped into above-ground reactor for separation of sludge and wastewater. As shown in **Figure 2**, the sludge upon separation is treated using BSFL into nutrient-rich frass applicable as fertilizer.

The wastewater upon pre-treatment is then pumped into open ponds containing microalgae for treatment. The accumulated biomass is harvested for further processing. The treated effluent water is tested for quality compliance before discharge or reuse. In general, microalgae are more suitable for batch processes which allow the

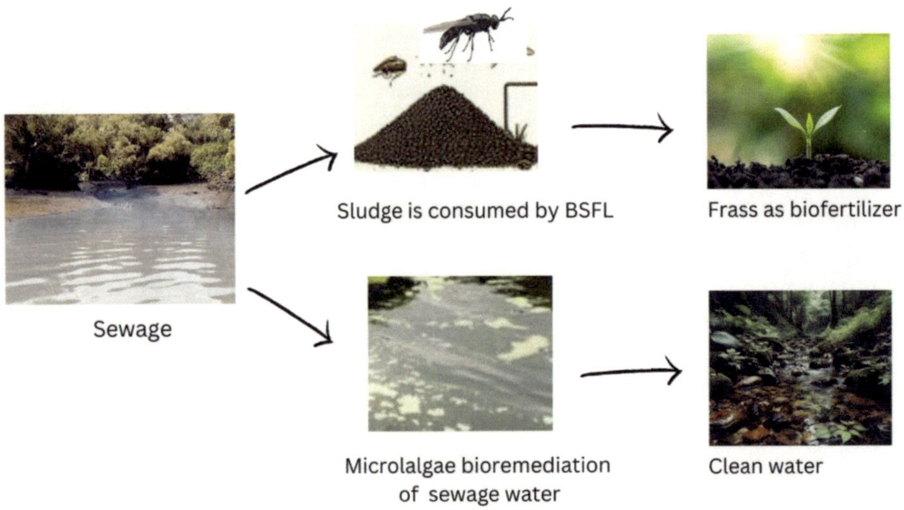

Figure 2.
Proposed set-up for sewage treatment using BSFL and microalagae.

recovery of biomass and treated effluent water. The method is sustainable, require minimal land use, easy to construct and maintain, require no energy input, and cheaper compared to conventional systems such as microfiltration. Similar to using BSFL, pumping costs may be offset by using solar energy or through cost sharing among households. Nevertheless, it should be noted that microalgae are more effective when the wastewater has already undergone pre-treatment. However, bacterial-algal consortia can be applied directly without sewage pre-treatment.

5. Future perspectives for sewage management using BSFL and microalgae

BSFL shows great potential in the decentralized treatment of sewage for small-scale settings. The method is still in its infancy and will need further investigation to establish optimal process conditions such as pumping rates, mixing ratios, agitation speed, aeration rate, and feed rate. Pilot studies will be vital in order to determine appropriate reactor designs and configurations. Since high FSR rates are achieved with the addition of nutrients or more economically with nutrient-rich domestic and market wastes derived from fruits and vegetables, optimal substrate mixing ratios will need to be established under different reaction conditions. Finally, a cost-benefit analysis will be required to determine optimal feed processing volumes and operational parameters that will guarantee recovery of investment and running costs.

Microalgae are more tailored for secondary and tertiary treatment of wastewater but not fecal sludge. A broader system will involve the use of bacterial-algal consortia. However, there is limited data for their use in low- and middle-income economies. Hence, there are opportunities for further research. BSFL and microalgae have shown potential when used separately for sewage treatment in these countries. In order to leverage their strengths, there is a need to investigate the performance and efficiency of the two systems when combined. BSFL will be vital for sludge and pre-treatment of wastewater. Micro-algae and bacteria-algal consortia will be useful for further treatment of wastewater to meet environmental quality limits.

6. Conclusions

This chapter has shown the potential, suitability, and sustainability of BSFL and microalgae in sewage management and treatment, particularly for small development settings employing sewerage systems based on cesspools and septic tanks. BSFL is more appropriate for sewage sludge digestion into nutrient-rich frass suitable for use as animal feed. Microalgae are more relevant to wastewater treatment allowing effluent reuse. Taking into consideration the sewerage systems and community settings in sub-Saharan African countries, significant benefits can be attained by combining BSFL and microalgae. Hence, above-ground treatment facilities employing pump and treat technologies need to be established within the settlement area. In order to minimize transportation costs, the sewage will need to be pumped, collected, and separated into sludge and wastewater. The sludge is fed into a suitable reactor containing BSFL while the wastewater is channeled into open ponds containing microalgae and bacteria-algal consortia. The reactors and open ponds should be designed to allow recovery of the products: frass and effluent water. Pumping costs can be minimized by the sale of products, for example, frass, cost sharing among several households, or more appropriately through utilization of solar energy. In summary, this chapter presents BSFL and microalgae as robust, cost-effective, and easy to establish, operate, and maintain technologies suitable for sewage treatment taking into consideration the common sewerage systems found in low-income communities and informal development settings in sub-Saharan Africa.

Author details

Ali Shee*, Mariam Maku Swaleh, Laila Uweso Abubakar, Huxley Mae Makonde, Saeed Mutta Mwaguni, Daniel Munga, Mumini Ndenge Dzoga and Cyrilus Otulo Wandera
Technical University of Mombasa, Mombasa, Kenya

*Address all correspondence to: sheeali@tum.ac.ke

References

[1] Guengant J-P. Africa's population: History, current status, and projections. In: Groth H, May JF, editors. Africa's Population: In Search of a Demographic Dividend. Cham: Springer International Publishing; 2017. pp. 11-31. DOI: 10.1007/978-3-319-46889-1_2

[2] United Nations. World Population Prospects. Geneva: Switzerland; 2017

[3] Linard C, Gilbert M, Snow RW, Noor AM, Tatem AJ. Population distribution, settlement patterns and accessibility across Africa in 2010. PLoS ONE. 2012;**7**:1-8. DOI: 10.1371/journal.pone.0031743

[4] OECD/SWAC. Africa's urbanisation dynamics 2020: Africapolis, mapping a new urban geography. In: West African Studies. Paris: OECD Publishing; 2020. DOI: 10.1787/b6bccb81-en

[5] OECD/WHO. Health at a Glance: Asia/Pacific 2020: Measuring Progress Towards Universal Health Coverage. Paris: OECD Publishing; 2020. DOI: 10.1787/f6a5d600-en

[6] Nansubuga I, Banadda N, Verstraete W, Rabaey K. A review of sustainable sanitation systems in Africa. Reviews in Environmental Science and Biotechnology. 2016;**15**:465-478. DOI: 10.1007/s11157-016-9400-3

[7] Lerebours A, Scott R, Sansom K, Kayaga S. Regulating sanitation services in Sub-Saharan Africa: An overview of the regulation of emptying and transport of faecal sludge in 20 cities and its implementation. Utilities Policy. 2021;**73**:101315. DOI: 10.1016/j.jup.2021.101315

[8] Marks RF. Appropriate sanitation options for southern Africa. Water Science and Technology. 1993;**27**:1-10. DOI: 10.2166/wst.1993.0003

[9] Munamati M, Nhapi I, Misi S. Exploring the determinants of sanitation success in Sub-Saharan Africa. Water Research. 2016;**103**:435-443. DOI: 10.1016/j.watres.2016.07.030

[10] Mwaguni S, Ayiemba EHO, Onyari JM. On-site sanitation and its effects on the groundwater resources of Nyali-Bamburi-Shanzu and Diani-Chale, Kenya. Journal of Environment and Earth Science. 2017;**7**:47-57

[11] Novotný J, Humňalová H, Kolomazníková J. The social and political construction of latrines in rural Ethiopia. Journal of Rural Studies. 2018;**63**:157-167. DOI: 10.1016/j.jrurstud.2018.08.003

[12] Dabwiso DF. Mitigating the Effects of Septic Tank Effluents from Households on Groundwater Quality: A Case of Meanwood-Kwamwena. Lusaka: University of Zambia; 2019

[13] Snyman HG. Management of wastewater and faecal sludge in southern Africa. Water Practice Technology. 2007;**2**:wpt2007089. DOI: 10.2166/wpt.2007.089

[14] Mekonnen A, Leta S, Njau KN. Wastewater treatment performance efficiency of constructed wetlands in African countries: A review. Water Science and Technology. 2014;**71**:1-8. DOI: 10.2166/wst.2014.483

[15] Onyango PO. Sanitation context and technological challenges to municipal wastewater management in Africa. In: Debik E, Bahadir M, Haarstrick A, editors. Wastewater Management and Technologies. Switzerland, Cham:

Springer Nature; 2023. pp. 183-191. DOI: 10.1007/978-3-031-36298-9_13

[16] Wang H, Wang T, Zhang B, Li F, Toure B, Omosa IB, et al. Water and wastewater treatment in Africa—Current practices and challenges. CLEAN - Soil, Air, Water. 2014;**42**:1029-1035. DOI: 10.1002/clen.201300208

[17] Okuku EO, Ohowa B, Mwangi SN, Munga D, Kiteresi LI, Wanjeri VO, et al. Sewage pollution in the coastal waters of Mombasa City, Kenya: A norm rather than an exception. International Journal of Environmental Research. 2011;**5**:865-874. DOI: 10.22059/ijer.2011.444

[18] Onu MA, Ayeleru OO, Oboirien B, Olubambi PA. Challenges of wastewater generation and management in sub-Saharan Africa: A review. Environmental Challenges. 2023;**11**:100686. DOI: 10.1016/j.envc.2023.100686

[19] Iannelli R, Giraldi D. Sources and composition of sewage effluent; treatment systems and methods. In: Treated Wastewater in Agriculture. United Kingdom: Blackwell Publishing Ltd; 2010. pp. 1-50. DOI: 10.1002/9781444328561.ch1

[20] Richardson SD, Kimura SY. Water analysis: Emerging contaminants and current issues. Analytical Chemistry. 2016;**88**:546-582. DOI: 10.1021/acs.analchem.5b04493

[21] K'oreje KO, Vergeynst L, Ombaka D, Wispelaere PD, Okoth M, Langenhove HV, et al. Occurrence patterns of pharmaceutical residues in wastewater, surface water and groundwater of Nairobi and Kisumu city, Kenya. Chemosphere. 2016;**149**:238-244. DOI: 10.1016/j.chemosphere.2016.01.095

[22] Fang YR, Li S, Zhang Y, Xie GH. Spatio-temporal distribution of sewage sludge, its methane production potential, and a greenhouse gas emissions analysis. Journal of Cleaner Production. 2019;**238**:117895. DOI: 10.1016/j.jclepro.2019.117895

[23] Foley J, Yuan Z, Lant P. Dissolved methane in rising main sewer systems: Field measurements and simple model development for estimating greenhouse gas emissions. Water Science and Technology. 2009;**60**:2963-2971. DOI: 10.2166/wst.2009.718

[24] WHO. Guidelines for the Safe Use of Wastewater, Excreta and Greywater. Vol. 4. Geneva: WHO; 2013

[25] Bakare BF. Innovative approach for the management of faecal sludge accumulated in ventilated improved pit latrine: A case study of eThekwini municipality in Durban South Africa. Journal of Physics Conference Series. 2019;**1378**:042006. DOI: 10.1088/1742-6596/1378/4/042006

[26] Saxena S, Den W. In situ treatment technologies for pit latrines to mitigate groundwater contamination by fecal pathogens: A review of recent technical advances. Journal of Water Sanitation and Hygiene for Development. 2021;**12**:102-115. DOI: 10.2166/washdev.2021.184

[27] Lee TA. Decentralized Sanitation Systems for Densely Populated Regions: Design, Prototyping, and Systems Value Analysis. Massachusetts, United States of America: Massachusetts Institute of Technology; 2021

[28] Olanrewaju D. Soak-away systems and possible groundwater pollution problems in developing countries. Journal of the Royal Society of Health. 1990;**110**:108-112. DOI: 10.1177/146642409011000314

[29] Sahoo KC, Nanda P, Prakash B, Lawande P, Sasidharan S, Nag D, et al. Landscape of sewerage and septage set-up in Puri, Odisha. In: Ghosh SK, editor. Sustainable Waste Management: Policies and Case Studies. Singapore: Springer Singapore; 2020. pp. 547-560

[30] Haydar S, Anis M, Hina GE, Aziz JA, Arbi M. An innovative design of septic tank for wastewater treatment and its performance evaluation: An applicable model for developing countries. Chinese Journal of Chemical Engineering. 2018;**26**:886-891. DOI: 10.1016/j.cjche.2017.10.015

[31] Singh RP, Kun W, Fu D. Designing process and operational effect of modified septic tank for the pre-treatment of rural domestic sewage. Journal of Environmental Management. 2019;**251**:109552. DOI: 10.1016/j.jenvman.2019.109552

[32] Nivala J, Headley T, Wallace S, Bernhard K, Brix H, van Afferden M, et al. Comparative analysis of constructed wetlands: The design and construction of the ecotechnology research facility in Langenreichenbach, Germany. Plants in Constructed, Restored and Created Wetlands. 2013;**61**:527-543. DOI: 10.1016/j.ecoleng.2013.01.035

[33] Shukla A, Parde D, Gupta V, Vijay R, Kumar R. A review on effective design processes of constructed wetlands. International Journal of Environmental Science and Technology. 2022;**19**:12749-12774. DOI: 10.1007/s13762-021-03549-y

[34] Cheng P, Jin Q, Jiang H, Hua M, Ye Z. Efficiency assessment of rural domestic sewage treatment facilities by a slacked-based DEA model. Journal of Cleaner Production. 2020;**267**:122111. DOI: 10.1016/j.jclepro.2020.122111

[35] Pan D, Hong W, Kong F. Efficiency evaluation of urban wastewater treatment: Evidence from 113 cities in the Yangtze River Economic Belt of China. Journal of Environmental Management. 2020;**270**:110940. DOI: 10.1016/j.jenvman.2020.110940

[36] USEPA. Guidelines for Water Reuse (No. EPA/600/R-12/618). Washington, D.C.; 2012

[37] Suryati T, Julaeha E, Farabi K, Ambarsari H, Hidayat AT. Lauric acid from the black soldier fly (Hermetia illucens) and its potential applications. Sustainability. 2023;**15**. DOI: 10.3390/su151310383

[38] Ferrarezi R, Cannella L, Nassef A, Bailey D. UVI/AES Annual Report 2016 - Alternative Sources of Food for Aquaponics in the U.S. Virgin Islands: A Case Study with Black Soldier Flies. United States of America: University of the Virgin Island Agricultural Experiment Station; 2016

[39] Diener S, Studt Solano NM, Roa Gutiérrez F, Zurbrügg C, Tockner K. Biological treatment of municipal organic waste using black soldier fly larvae. Waste and Biomass Valorization. 2011;**2**:357-363. DOI: 10.1007/s12649-011-9079-1

[40] Liu T, Awasthi MK, Chen H, Duan Y, Awasthi SK, Zhang Z. Performance of black soldier fly larvae (Diptera: Stratiomyidae) for manure composting and production of cleaner compost. Journal of Environmental Management. 2019;**251**:109593. DOI: 10.1016/j.jenvman.2019.109593

[41] Popa R, Green TR. Using black soldier fly larvae for processing organic leachates. Journal of Economic Entomology. 2012;**105**:374-378. DOI: 10.1603/EC11192

[42] Tschirner M, Simon A. Influence of different growing substrates and

processing on the nutrient composition of black soldier fly larvae destined for animal feed. Journal of Insects as Food and Feed. 2015;**1**:249-259. DOI: 10.3920/JIFF2014.0008

[43] Singh A, Kumari K. An inclusive approach for organic waste treatment and valorisation using Black Soldier Fly larvae: A review. Journal of Environmental Management. 2019;**251**:109569. DOI: 10.1016/j.jenvman.2019.109569

[44] Achieng Oyoo V, Riungu JN, Dey P, Kirimi JG, Matheka RM. Process performance evaluation of faecal matter treatment via black soldier fly. Journal of Water Sanitation and Hygiene for Development. 2023;**13**:441-452. DOI: 10.2166/washdev.2023.010

[45] Nyakeri EM, Ayieko MA, Amimo FA, Salum H, Ogola HJO. An optimal feeding strategy for black soldier fly larvae biomass production and faecal sludge reduction. Journal of Insects as Food and Feed. 2019;**5**:201-213. DOI: 10.3920/JIFF2018.0017

[46] Lalander C, Diener S, Magri ME, Zurbrügg C, Lindström A, Vinnerås B. Faecal sludge management with the larvae of the black soldier fly (Hermetia illucens)—From a hygiene aspect. Science of the Total Environment. 2013;**458-460**:312-318. DOI: 10.1016/j.scitotenv.2013.04.033

[47] Banks IJ, Gibson WT, Cameron MM. Growth rates of black soldier fly larvae fed on fresh human faeces and their implication for improving sanitation. Tropical Medicine & International Health. 2014;**19**:14-22. DOI: 10.1111/tmi.12228

[48] Tokwaro R, Semiyaga S, Niwagaba CB, Nakagiri A, Sempewo JI, Muoghalu CC, et al. Application of black soldier fly larvae in decentralized treatment of faecal sludge from pit latrines in informal settlements in Kampala city. Frontiers in Environmental Science. 2023;**11**:1-11. DOI: 10.3389/fenvs.2023.1118635

[49] Cai M, Hu R, Zhang K, Ma S, Zheng L, Yu Z, et al. Resistance of black soldier fly (Diptera: Stratiomyidae) larvae to combined heavy metals and potential application in municipal sewage sludge treatment. Environmental Science and Pollution Research. 2018;**25**:1559-1567. DOI: 10.1007/s11356-017-0541-x

[50] Raksasat R, Abdelfattah EA, Liew CS, Rawindran H, Kiatkittipong K, Mohamad M, et al. Enriched sewage sludge from anaerobic pre-treatment in spurring valorization potential of black soldier fly larvae. Environmental Research. 2022;**212**:113447. DOI: 10.1016/j.envres.2022.113447

[51] Arnone S, De Mei M, Petrazzuolo F, Musmeci S, Tonelli L, Salvicchi A, et al. Black soldier fly (*Hermetia illucens* L.) as a high-potential agent for bioconversion of municipal primary sewage sludge. Environmental Science and Pollution Research. 2022;**29**:64886-64901. DOI: 10.1007/s11356-022-20250-w

[52] Barroca MJ, Florença SG, Martins CB, Guiné RPF. Chapter 48—Strategy and design of innovation policy road mapping for microalgae-based foods. In: Jacob-Lopes E, Queiroz MI, Maroneze MM, Zepka LQ, editors. Handbook of Food and Feed from Microalgae. United States of America: Academic Press; 2023. pp. 625-642. DOI: 10.1016/B978-0-323-99196-4.00045-0

[53] Levasseur W, Perré P, Pozzobon V. A review of high value-added molecules production by microalgae in light of the classification. Biotechnology Advances.

2020;**41**:107545. DOI: 10.1016/j.biotechadv.2020.107545

[54] Silva J, Alves C, Pinteus S, Reboleira J, Pedrosa R, Bernardino S. Chapter 3.10—Chlorella. In: Nabavi SM, Silva AS, editors. Nonvitamin and Nonmineral Nutritional Supplements. United States of America: Academic Press; 2019. pp. 187-193. DOI: 10.1016/B978-0-12-812491-8.00026-6

[55] Singh P, Gupta SK, Guldhe A, Rawat I, Bux F. Chapter 4—Microalgae isolation and basic culturing techniques. In: Kim S-K, editor. Handbook of Marine Microalgae. Boston: Academic Press; 2015. pp. 43-54. DOI: 10.1016/B978-0-12-800776-1.00004-2

[56] Armbrust EV. The life of diatoms in the world's oceans. Nature. 2009;**459**:185-192. DOI: 10.1038/nature08057

[57] Swaleh M, Abubakar L, Mwaguni S, Munga D, Okuku E. The potential of marine micro-algae grown in wastewater to remove nutrients and produce biomass. The Kenya Aquatica Journal. 2022a;**7**:22-32

[58] Parus A, Karbowska B. Marine algae as natural indicator of environmental cleanliness. Water, Air, and Soil Pollution. 2020;**231**:97. DOI: 10.1007/s11270-020-4434-0

[59] Chowdury MK, Nahar N, Deb UK. The growth factors involved in microalgae cultivation for biofuel production: A review. Computational Water, Energy, and Environmental Engineering. 2020;**9**:185-215. DOI: 10.4236/cweee.2020.94012

[60] Juneja A, Ceballos RM, Murthy GS. Effects of environmental factors and nutrient availability on the biochemical composition of algae for biofuels production: A review. Energies. 2013;**6**:4607-4638. DOI: 10.3390/en6094607

[61] Swaleh M, Abubakar L, Mwaguni S, Munga D, Okuku E, Dzoga M, et al. Effect of selected environmental factors on microalgae diversity and abundance in Gazi Bay, south coast Kenya. Journal of Sea Research. 2022b;**184**:102217. DOI: 10.1016/j.seares.2022.102217

[62] Geremia E, Ripa M, Catone CM, Ulgiati S. A review about microalgae wastewater treatment for bioremediation and biomass production—A new challenge for Europe. Environments. 2021;**8**:1-24. DOI: 10.3390/environments8120136

[63] Herold C, Ishika T, Nwoba EG, Tait S, Ward A, Moheimani NR. Biomass production of marine microalga *Tetraselmis suecica* using biogas and wastewater as nutrients. Biomass and Bioenergy. 2021;**145**:105945. DOI: 10.1016/j.biombioe.2020.105945

[64] Craggs RJ, McAuley PJ, Smith VJ. Wastewater nutrient removal by marine microalgae grown on a corrugated raceway. Water Research. 1997;**31**:1701-1707. DOI: 10.1016/S0043-1354(96)00093-0

[65] Jiang L, Luo S, Fan X, Yang Z, Guo R. Biomass and lipid production of marine microalgae using municipal wastewater and high concentration of CO_2. Spec. Issue Energy Algae Curr. Status Future Trends. Applied Energy. 2011;**88**:3336-3341. DOI: 10.1016/j.apenergy.2011.03.043

[66] Abdelfattah A, Ali SS, Ramadan H, El-Aswar EI, Eltawab R, Ho S-H, et al. Microalgae-based wastewater treatment: Mechanisms, challenges, recent advances, and future prospects. Environmental Science and

Ecotechnology. 2023;**13**:100205. DOI: 10.1016/j.ese.2022.100205

[67] López Rosales L, López-García P, Benyachou MA, Molina-Miras A, Gallardo J, Cerón-García MC, et al. Treatment of secondary urban wastewater with a low ammonium-tolerant marine microalga using zeolite-based adsorption. Bioresource Technology. 2022;**359**:127490. DOI: 10.1016/j.biortech.2022.127490

[68] Zhou W, Wang J, Chen P, Ji C, Kang Q, Lu B, et al. Bio-mitigation of carbon dioxide using microalgal systems: Advances and perspectives. Renewable and Sustainable Energy Reviews. 2017;**76**:1163-1175. DOI: 10.1016/j.rser.2017.03.065

[69] Wang M, Ye X, Bi H, Shen Z. Microalgae biofuels: Illuminating the path to a sustainable future amidst challenges and opportunities. Biotechnology for Biofuels and Bioproducts. 2024;**17**:10. DOI: 10.1186/s13068-024-02461-0

[70] Costa JAV, Freitas BCB, Santos TD, Mitchell BG, Morais MG. Chapter 9—Open pond systems for microalgal culture. In: Pandey A, Chang J-S, Soccol CR, Lee D-J, Chisti Y, editors. Biofuels from Algae. 2nd ed. Amsterdam, Netherlands: Elsevier; 2019. pp. 199-223. DOI: 10.1016/B978-0-444-64192-2.00009-3

[71] Tan JS, Lee SY, Chew KW, Lam MK, Lim JW, Ho S-H, et al. A review on microalgae cultivation and harvesting, and their biomass extraction processing using ionic liquids. Bioengineered. 2020;**11**:116-129. DOI: 10.1080/21655979.2020.1711626

[72] Lee Y-J, Lei Z. Microalgal-bacterial aggregates for wastewater treatment: A mini-review. Bioresource Technology Reports. 2019;**8**:100199. DOI: 10.1016/j.biteb.2019.100199

[73] Hajri AK, Alsharif I, Albalawi MA, Alshareef SA, Albalawi RK, Jamoussi B. Utilizing mixed cultures of microalgae to up-cycle and remove nutrients from dairy wastewater. Biology. 2024;**13**. DOI: 10.3390/biology13080591

[74] Aditya L, Mahlia TMI, Nguyen LN, Vu HP, Nghiem LD. Microalgae-bacteria consortium for wastewater treatment and biomass production. Science of the Total Environment. 2022;**838**:155871. DOI: 10.1016/j.scitotenv.2022.155871

[75] Nandeshwar S, Satpute G. Green technical methods for treatment of waste water using microalgae and its application in the management of natural water resources: A review. Current World Environment. 2014;**9**:837-842. DOI: 10.12944/CWE.9.3.34

[76] Tang C-C, Hu Y-R, He Z-W, Li Z-H, Tian Y, Wang XC. Promoting symbiotic relationship between microalgae and bacteria in wastewater treatment processes: Technic comparison, microbial analysis, and future perspectives. Chemical Engineering Journal. 2024;**498**:155703. DOI: 10.1016/j.cej.2024.155703

[77] Rossi S, Pizzera A, Bellucci M, Marazzi F, Mezzanotte V, Parati K, et al. Piggery wastewater treatment with algae-bacteria consortia: Pilot-scale validation and techno-economic evaluation at farm level. Bioresource Technology. 2022;**351**:127051. DOI: 10.1016/j.biortech.2022.127051

[78] Singh B, Sharma N, Sharma SG, Dhir A, Kocher GS, Mamane H. Application of microbial consortium to the pilot-scale treatment of primary treated sewage wastewater. Water, Air,

& Soil Pollution. 2024;**235**:132. DOI: 10.1007/s11270-024-06953-5

[79] Bhatia KS, Ahuja V, Chandel N, Mehariya S, Kumar P, Vinayak V, et al. An overview on microalgal-bacterial granular consortia for resource recovery and wastewater treatment. Bioresource Technology. 2022;**351**:127028. DOI: 10.1016/j.biortech.2022.127028

[80] Ren Z, Li H, Sun P, Fu R, Bai Z, Zhang G, et al. Development and challenges of emerging biological technologies for algal-bacterial symbiosis systems: A review. Bioresource Technology. 2024;**413**:131459. DOI: 10.1016/j.biortech.2024.131459

[81] Jiang Q, Chen H, Fu Z, Fu X, Wang J, Liang Y, et al. Current progress, challenges and perspectives in the microalgal-bacterial aerobic granular sludge process: A review. International Journal of Environmental Research and Public Health 19. 2022. DOI: 10.3390/ijerph192113950

[82] Walls LE, Velasquez-Orta SB, Romero-Frasca E, Leary P, Yáñez Noguez I, Orta Ledesma MT. Non-sterile heterotrophic cultivation of native wastewater yeast and microalgae for integrated municipal wastewater treatment and bioethanol production. Biochemical Engineering Journal. 2019;**151**:107319. DOI: 10.1016/j.bej.2019.107319

[83] Chen M, Chang L, Zhang J, Guo F, Vymazal J, He Q, et al. Global nitrogen input on wetland ecosystem: The driving mechanism of soil labile carbon and nitrogen on greenhouse gas emissions. Environmental Science and Ecotechnology. 2020;**4**:100063. DOI: 10.1016/j.ese.2020.100063

[84] Nguyen T-T-D, Nguyen T-T, An Binh Q, Bui X-T, Ngo HH, Vo HNP, et al. Co-culture of microalgae-activated sludge for wastewater treatment and biomass production: Exploring their role under different inoculation ratios. Bioresource Technology. 2020;**314**:123754. DOI: 10.1016/j.biortech.2020.123754

[85] Mujtaba G, Rizwan M, Kim G, Lee K. Removal of nutrients and COD through co-culturing activated sludge and immobilized Chlorella vulgaris. Chemical Engineering Journal. 2018;**343**:155-162. DOI: 10.1016/j.cej.2018.03.007

[86] Mujtaba G, Lee K. Treatment of real wastewater using co-culture of immobilized Chlorella vulgaris and suspended activated sludge. Water Research. 2017;**120**:174-184. DOI: 10.1016/j.watres.2017.04.078

[87] El-Aswar EI, Ramadan H, Elkik H, Taha AG. A comprehensive review on preparation, functionalization and recent applications of nanofiber membranes in wastewater treatment. Journal of Environmental Management. 2022;**301**:113908. DOI: 10.1016/j.jenvman.2021.113908

[88] Eroglu E, Agarwal V, Bradshaw M, Chen X, Smith SM, Raston CL, et al. Nitrate removal from liquid effluents using microalgae immobilized on chitosan nanofiber mats. Green Chemistry. 2012;**14**:2682-2685. DOI: 10.1039/C2GC35970G

[89] Vasistha S, Khanra A, Rai MP. Influence of microalgae-ZnO nanoparticle association on sewage wastewater towards efficient nutrient removal and improved biodiesel application: An integrated approach. Journal of Water Process Engineering. 2021;**39**:101711. DOI: 10.1016/j.jwpe.2020.101711

Chapter 3

Treatment and Recovery of Biosolids from Sewage Wastewater Using Bioflocculants *Bacillus velezensis* Isolate JB7

Hassimi Abu Hasan, Hema Darshinni Vijayakumar, Jahira Alias, Nur 'Izzati Ismail and Setyo Budi Kurniawan

Abstract

Sewage wastewater contains pollutants that may pollute the environment if not properly treated. The wastewater also contains solids that can be recovered and used as biofertilizer, supporting the circular economy concept. Bioflocculant, a biopolymer produced by microorganisms, is considered an effective and environmentally friendly wastewater treatment method. The aim of this study was to evaluate the effect of bioflocculant dosage, type of cation addition, and mixing speed on the treatment and recovery of biosolids from sewage wastewater. The addition of bioflocculant produced by *Bacillus velezensis* isolate JB7 in the test varied from 1 to 3 mL per 450 mL of sewage wastewater. Three types of cations were studied such as Al^{3+} (from aluminum sulfate [$Al_2(SO4)_3$]), Fe^{3+}(from ferric sulfate [$Fe_2(SO_4)_3$]), and Ca^{2+} (from calcium chloride [$CaCl_2$]), while the rapid mixing speed was varied from 100 to 200 rpm. The results show that the performance of COD, NH_3-N, and biosolid recovery varied significantly based on the type of cation used. The highest COD, NH_3-N, and biosolid recovery were achieved with a dosage of 2 mL using $Al_2(SO_4)_3$, $CaCl_2$, and $Fe_2(SO_4)_3$, respectively. The recovery of biosolids from sewage wastewater is a beneficial bio-product that can be used as fertilizer. However, detailed studies on the potential of these biosolids need to be further investigated in the future.

Keywords: sewage wastewater, circular economy, biofertilizer, bioflocculant, effluent quality

1. Introduction

Sewage wastewater comprises substances including organic and inorganic matter, heavy metals, pathogens, nutrients, and contaminants of emerging concern (CECs). It is generated from various sources, such as domestic households (e.g., bathing, cooking, washing, and toilet flushing) and commercial effluents (e.g., restaurants, laundries, and hospitals) [1]. Untreated or improperly treated sewage wastewater significantly harms the environment and human health due to water pollution, which deteriorates water quality and disrupts the balance of aquatic life [2]. Additionally, various problems may occur, such as eutrophication, algal blooms, waterborne diseases, and soil contamination. Therefore, sewage wastewater should be treated to reduce the transmission of infections, lessen water contamination, and minimize the subsequent harm to aquatic life.

Sewage wastewater requires treatment before being released or, ideally, reused for agricultural irrigation. Interestingly, sewage wastewater contains value-added products that can be used as renewable and sustainable resources, such as biosolids and bio-struvite. Technologies for sewage wastewater recovery include coagulation-flocculation, aerobic-anaerobic systems, extended aeration, membranes, ammonia stripping, and electrochemical methods [2, 3]. Researchers are increasingly focusing on using cleaner and greener technologies for the treatment and recovery of biosolids from sewage, particularly through the use of bioflocculants in the flocculation process.

2. Bioflocculants and biosolids

Bioflocculants are extracellular biopolymers that contain proteins, polysaccharides, lipids, cellulose, nucleic acids, and glycolipids, produced by microorganisms such as bacteria, fungi, and algae. These natural compounds are biodegradable, non-toxic, and environmentally friendly compared to chemical flocculants [4]. The composition of bioflocculants varies depending on the culture medium, operating culture conditions, and microorganisms used. They have the ability to flocculate suspended solids and colloidal particles into larger flocs. These larger flocs can be easily separated from aqueous solutions through methods such as sedimentation, filtration, or flotation. Bioflocculants exhibit high flocculation efficiency and pollutant removal effectiveness comparable to chemical flocculants [5, 6]. Bioflocculant facilitate the agglomeration of suspended solids, including organic materials, in wastewater via several mechanisms including charge neutralization, sweep flocculation, bridging, or patch flocculation [7, 8] to produce biosolids.

Biosolids or biological sludge produced from treatment using bioflocculants are highly biodegradable and rich in nutrients, making them suitable as soil conditioners or fertilizers. They can also act as plant growth boosters and are referred to as slow-release fertilizers since they release nutrients gradually. Their nutrient content and plant growth efficiency are comparable to chemical fertilizers [8]. Biosolids primarily contain phosphorus and nitrogen and a variety of macronutrients (e.g., Mg, Ca, S, P, and N) and micronutrients (e.g., Mo, Zn, Cu, and Mn). Considering concerns about fossil fuel depletion and climate change, using biosolids as fertilizer benefits the water industry and agriculture sector by promoting renewable resources for future generations. Additionally, biosolids can be utilized for liquid fuels and chemical production [9]. Given these benefits, the potential recovery of biosolids from sewage wastewater promotes a circular economy in the sector.

3. Sewage wastewater collection

Sewage wastewater was collected in a 5 L plastic container from the sewage treatment plant (STP) located in the student hostel Kolej Keris Mas at UKM. The STP treatment process includes screening, an extended aeration system, and sedimentation. The wastewater sampling point was located before the screening process of the STP (see **Figure 1**). After collection, it was stored in dark surroundings at room temperature of 4°C prior to use. **Figure 1** illustrates the sampling point of the sewage wastewater.

Figure 1.
Location of sewage wastewater collection.

4. Production of bioflocculant by *Bacillus velezensis* isolate JB7

Bioflocculant-producing bacteria (BPB) identified as *Bacillus velezensis* isolate JB7 (MW386826) was isolated from the Langat River, Selangor. The stock culture of JB7 was grown in nutrient broth (Oxoid, CM0001) in an incubator (Protech, Model S1-100D, Malaysia) at 37°C for 24 hours to be used as seed culture for bioflocculant production. Afterward, 10%v/v of the JB7 seed culture was transferred into 90 mL nutrient broth in conical flasks and incubated in a shaker incubator (Protech, Model S1-100D, Malaysia) at 37°C and 150 rpm for 24 hours. After 24 hours, the fermented broth was centrifuged (Centrifuge 5810 Eppendorf AG, Germany) at 4000 rpm for 15 minutes to separate the solids which consist of JB7 biomass from cell-free supernatant. The final product of this process is the bioflocculant, which consists of the cell-free supernatant. This bioflocculant is used for biosolid recovery from sewage wastewater in the next process.

5. Biosolid recovery from sewage using bioflocculant

The testing of biosolid recovery in the sewage wastewater was conducted using a flocculator (VELP Scientifica, model JLT6). Three factors were explored to recover the biosolid from sewage (**Table 1**) which are dosage of bioflocculant (1–3 mL per 450 mL sewage), mixing speed of the flocculator (100–200 rpm), and type of cation given by chemical compounds (aluminum sulfate, ferric sulfate, and calcium chloride). For the effect of bioflocculant dosage, 1–3 mL of bioflocculant was added to each beaker containing 450 mL wastewater sample. The mixer of the flocculator was turned on at a stirring speed of 100 rpm for 2 minutes to obtain a thorough dispersion of bioflocculants in the wastewater. After that, the stirring speed was reduced to 30 rpm for 20 minutes to stimulate the agglomeration process of the biosolid. The

Bioflocculant dosage (mL)	1–3
Mixing speed (rpm)	100–200
Type of cation	Aluminum sulfate, ferric sulfate, and calcium chloride

Table 1.
Jar test conditions using JB7 bioflocculants.

floc of biosolid in the beakers was allowed to settle for 30 minutes. For the effect of rapid mixing speed from 100 to 200 rpm, the bioflocculant volume was fixed at 2 mL without cation addition. Meanwhile, for the effect of cation type, the mixing speed was maintained at 100 rpm and bioflocculant volume was fixed at 2 mL.

6. Quality check of the sewage wastewater

Sewage wastewater parameter quality which consists of nitrite-nitrogen (NO_2^--N), nitrate-nitrogen (NO_3^--N), chemical oxygen demand (COD), ammoniacal-nitrogen (NH_3-N), total suspended solid (TSS), pH, color and turbidity were analyzed throughout this study. The NH_3-N was determined using the Nessler method (Method 8038), while NO_2^--N and NO_3^--N, and PO_4^{3-}-P were determined using Methods 8507, 8039, and 8048, respectively. For COD analysis, it was measured using Method 8000. About 2 mL of sample was pipetted into the COD vial and was shaken for few times. The vial was placed into the COD digestion for 2 hours at 120°C. After the digestion, the vial was cooled down prior to analysis. Meanwhile, TSS and color were analyzed using Methods 8006 and 8025, respectively. The measurement of the parameters was determined using HACH DR3900 spectrophotometer. All the procedures were described by the HACH manufacturer [10]. pH-Value was measured using pH meter (Metrohm 827, United States). In addition, turbidity was measured using turbidimeter (Hach, Model 2100 N, United States). The removal efficiency (*R*) of the COD and NH_3-N was measured using Eq. (1).

$$R = \frac{C_i - C_e}{C_i} x100 \qquad (1)$$

where C_i is the raw sewage wastewater concentration and C_e is the treated sewage wastewater for COD and NH_3-N parameters.

7. Profile of the collected sewage wastewater

The profile of collected sewage wastewater includes physical and chemical characteristics. Key parameters are pH, nutrient levels, and solid contents. **Table 2** presents the characterization of raw sewage wastewater sampled from the STP at KKM, UKM. The STP in which the collection work was done could handle a population equivalent (PE) of 2700 people. A PE of 2700 people represents the sewage wastewater load generated by 2700 individuals, typically measured in terms of organic and nutrient load, and flow rate. Low strength of sewage wastewater was observed due to low pollution loads. The pollutants in sewage primarily result from daily activities such as bathing, washing, and cooking. Low concentrations of pollutants are not unusual; this

Parameter	Value
pH	6.85
Total suspended solid, TSS (mg/L)	34
Total dissolved solid, TDS (mg/L)	242
Turbidity (NTU)	17.8
MLSS (mg/L)	31
COD (mg/L)	108
Ammonia-nitrogen (mg/L)	4.46
Nitrate (mg/L)	7.4
Nitrite (mg/L)	2.58
Phosphate (mg/L)	4.25
Color (unit PtCo)	180

Table 2.
Raw domestic wastewater characterization.

has been widely reported in the literature. For example, Li et al. [11] found pollutant concentrations in sewage to be 55.24 mg/L for COD, 0.34 mg/L for NH_3–N, 0.77 mg/L for total phosphorus (TP), and 3.32 mg/L for BOD_5.

8. Removal of chemical oxygen demand

COD is used to measure the amount of organic and inorganic substances in sewage wastewater that can be chemically oxidized. The effect of bioflocculant dosage on COD removal is shown in **Figure 2(a)**. Lower and higher COD removals indicate the efficiency of the bioflocculants in the treatment of sewage wastewater. The removal of COD increased from 29 to 54.9% as the bioflocculant dosage was increased to 2 mL. At a higher dosage of 3 mL, the COD removal decreased to 41.7%. A low dosage of bioflocculant leads to inefficient collisions and adsorption of organic particles that contribute to low COD removal. In contrast, a high dosage of bioflocculant creates stronger inter-repulsive forces among the bioflocculants, which hinders particle movement and reduces effective collisions [12]. The hindrance in particle movement reduces the formation of macroflocs, leading to less efficiency in the clarification process. Consequently, more suspended and dissolved organic compounds remain in the sewage wastewater.

Figure 2(b) shows the effect of cation addition along with bioflocculants in the sewage wastewater. The addition of trivalent ion Al^{3+} [$Al_2(SO_4)_3$] resulted in the highest COD removal at 54%, followed by $CaCl_2$ and $Fe_2(SO_4)_3$, which achieved COD removals of 51.2 and 46.6%, respectively. $Al_2(SO_4)_3$ is a strong coagulant where it has a high charge density to enhance the neutralization of negatively charged organic particles and colloids in sewage wastewater. $Fe_2(SO_4)_3$ is also a trivalent ion but this coagulant forms less stable flocs and is influenced by pH conditions, leading to a slightly lower COD removal compared to $Al_2(SO_4)_3$. In addition, $CaCl_2$ has lower charge density compared to $Al_2(SO_4)_3$. This coagulant reduces the COD through charge neutralization and bridging mechanisms. Agunbiade et al. [13] found that adding 1% $CaCl_2$ to sewage wastewater containing 0.8 mg/mL of bioflocculant, and agitating the mixture at 160 rpm for 2 minutes, resulted in a 65.7% removal of COD.

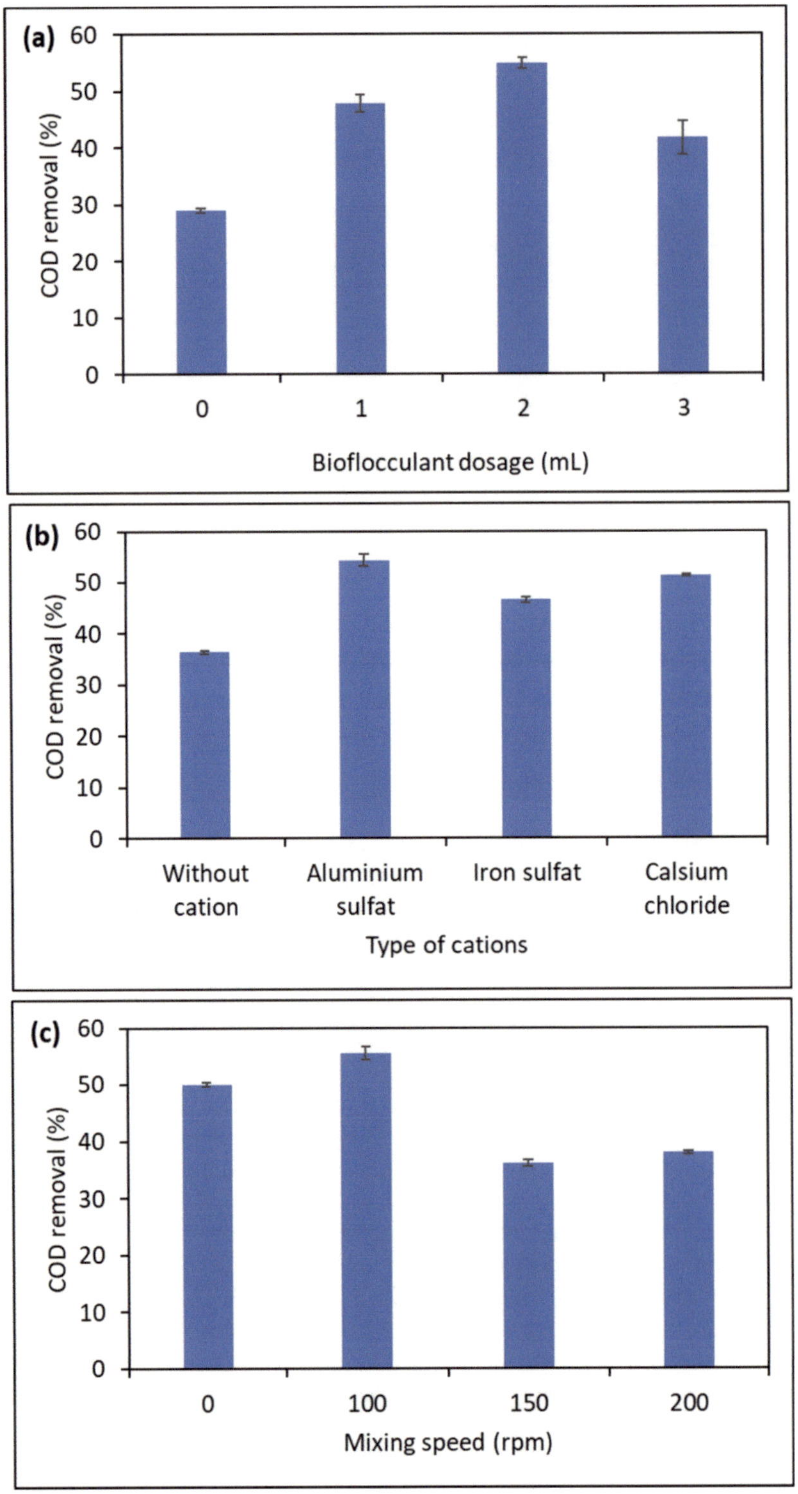

Figure 2.
Performance of chemical oxygen demand removal (a) effect of bioflocculant dosage, (b) effect of type of cation, and (d) effect of mixing speed.

Figure 2(c) presents the influence of mixing speed. The highest COD removal, 55.6%, was achieved at a mixing speed of 100 rpm. When the mixing speed was increased to 150 and 200 rpm, the COD removal decreased to 36.1 and 37.9%, respectively. The COD removal decreased at higher mixing speeds (150 and 200 rpm) due to the hindrance of particle aggregation and floc formation. The mixing speed at 100 rpm is an optimal condition for promoting collisions between particles and coagulants, allowing for the formation of stable flocs. However, at higher speeds, excessive mixing disrupts particle movement and damages formed flocs, resulting in lower settling efficiency and COD removal.

9. Removal of ammoniacal-nitrogen

Ammoniacal-nitrogen (NH_3-N) is a concentration of nitrogen present in sewage wastewater in the form of ammonia (NH_3) and ammonium ions (NH_4^+). A high value of NH_3-N discharge in the wastewater effluent indicates can lead to eutrophication. A low value in the discharge exhibits effective wastewater treatment plants, thus could protect and conserve the environment from pollution. The performance of NH_3-N removal under the effect of bioflocculant dosage is shown in **Figure 3(a)**. From the graph, increasing the bioflocculant dosage from 1 to 3 mL per 450 mL of sewage wastewater decreased the NH_3-N removal from 56.1 to 50%.

The decrease in NH_3-N removal indicates that excessive bioflocculant may disrupt the treatment process, leading to higher residual ammonia levels in the treated sewage wastewater. The results show that 1 mL of bioflocculant is the optimum dosage to achieve the highest removal efficiency. When the divalent calcium ion (Ca^{2+}) was added, it achieved the highest removal rate of ammonia-nitrogen (NH_3-N) at 40.4%. This was followed by the trivalent ions ferric (Fe^{3+}) and aluminum (Al^{3+}), as shown in **Figure 3(b)**. The results indicate that NH_3-N removal only requires cation-independent bioflocculants. According to Tang et al. [14], there were limited cation-independent bioflocculants have been reported. Cation-independent bioflocculants function through mechanisms such as adsorption, charge neutralization, and bridging, where the bioflocculant molecules directly interact with NH_3-N compounds without relying heavily on cationic ions like Ca^{2+}, Fe^{3+}, or Al^{3+}. The bioflocculants added to the sewage wastewater can bind to NH_3-N through the mechanisms, forming larger flocs, thereby reducing NH_3-N effectively. For the effect of rapid mixing speed (**Figure 3c**), increasing the speed from 100 to 200 rpm reduced the removal of NH_3-N. The highest removal was achieved at 100 rpm with removal of 22.9%, while the lower was at 200 rpm. At 100 rpm mixing speed, forms optimal floc and interaction with NH_3-N in the sewage wastewater, leading to higher removal efficiency. However, rapid mixing speeds at 200 rpm cause excessive turbulence, disrupting the formation and stability of flocs, and hindering the effective adsorption of NH_3-N compounds on the flocs formed. At a mixing speed of 350 rpm and using bioflocculant Pga21, Pan et al. [15] found that the removal of NH_3-N from the domestic wastewater was 33.2%.

10. Efficiency of biosolid recovery

Biosolid recovery was measured through the formation of a settle floc at the bottom of the beaker. Analysis shows that a bioflocculant dosage of 2 mL resulted in the highest biosolid recovery, with 31 mg/L, compared to dosages of 1 and 3 mL per 450 mL of sewage wastewater, which recovered only 27 and 19.3 mg/L, respectively (**Figure 4**).

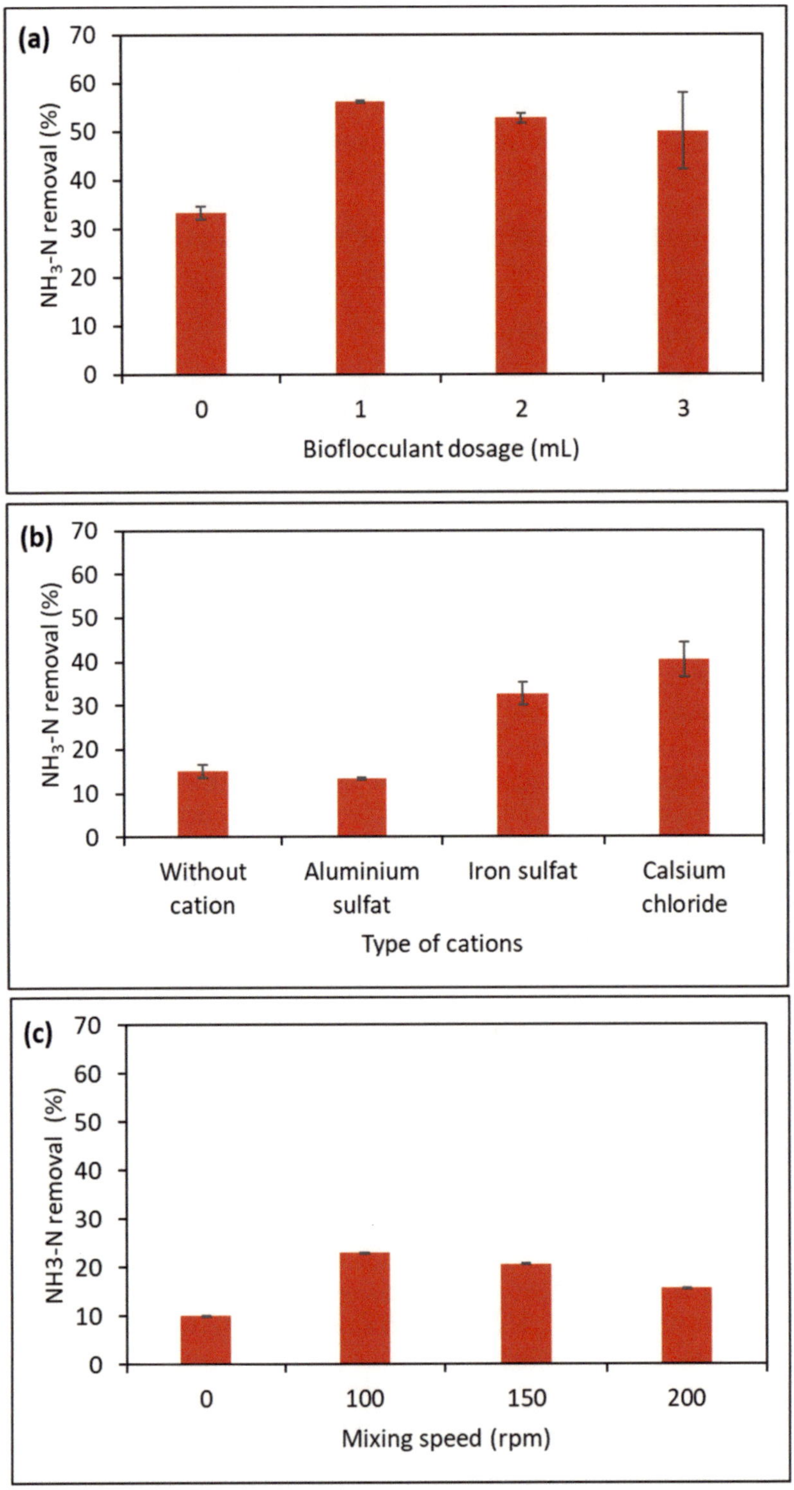

Figure 3.
Performance of ammoniacal-nitrogen removal (a) effect of bioflocculant dosage, (b) effect of type of cation, and (d) effect of mixing speed.

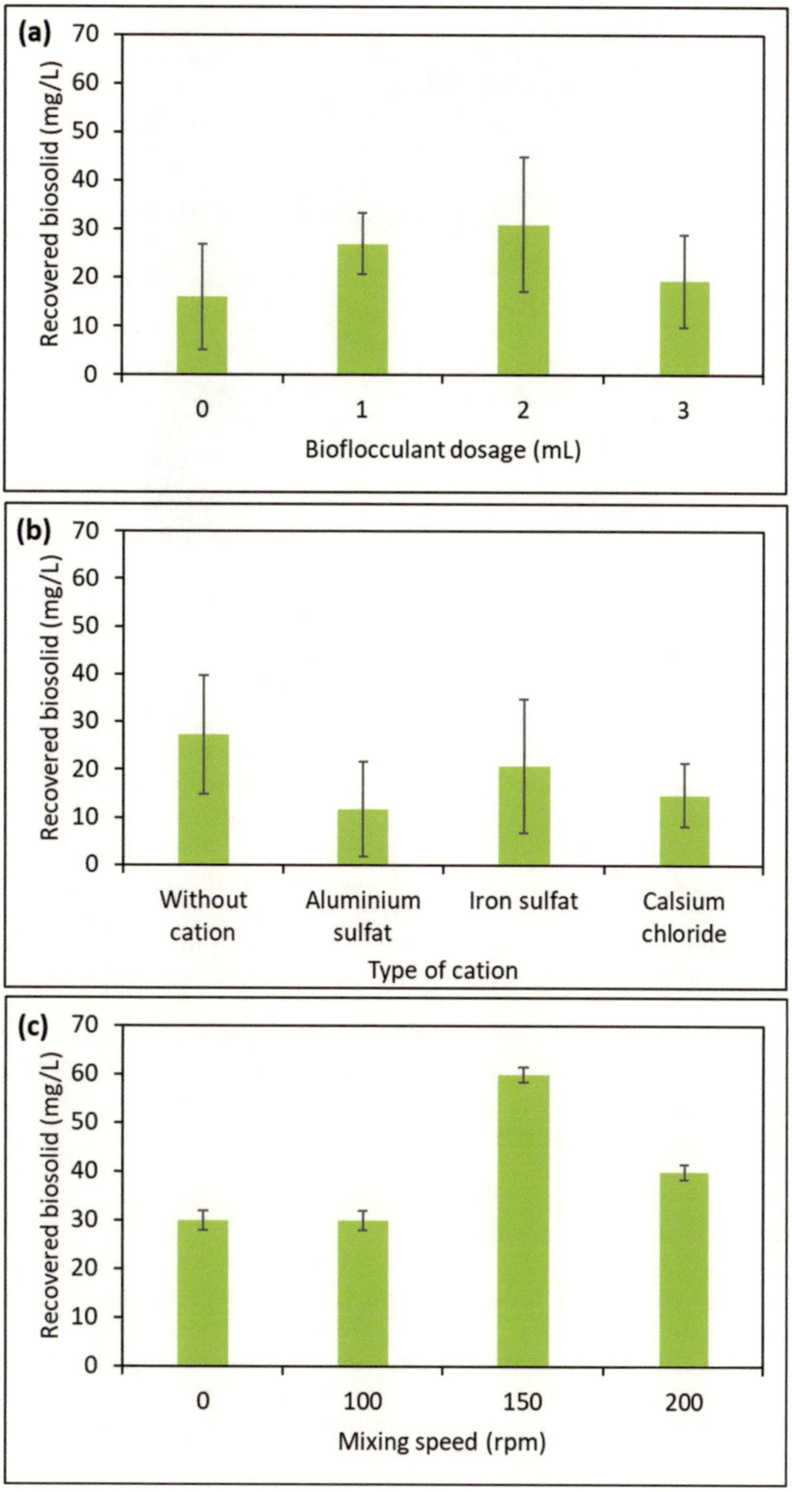

Figure 4.
Performance of ammoniacal-nitrogen removal (a) effect of bioflocculant dosage, (b) effect of type of cation, and (d) effect of mixing speed.

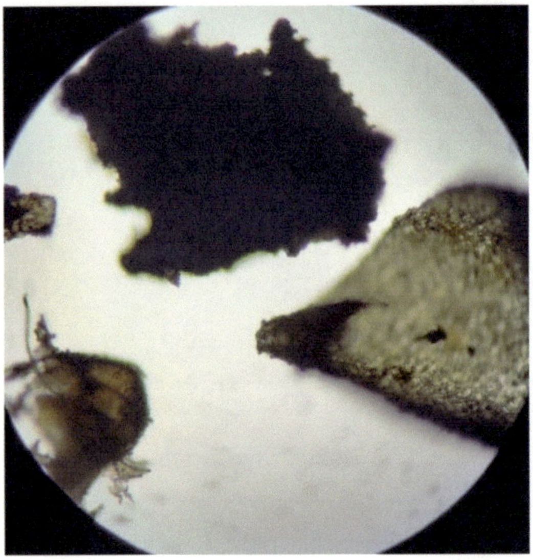

Figure 5.
Image of biosolid recovered from sewage wastewater at 10x magnification.

In comparison to adding cations such as $Al_2(SO_4)_3$, $CaCl_2$, and $Fe_2(SO_4)_3$, more biosolid could be recovered from the sewage wastewater without the addition of these components. This indicates that the bioflocculant does not interact with the cation to enhance the formation of solid agglomeration, suggesting the bioflocculant produced by JB7 is a cation-independent for solid recovery. This observation aligns with Tang et al. [14] who reported that the bioflocculant ETH-2 produced by *Enterobacter* sp., resulted in high flocculating activity without the need for cation supplementation.

Regarding mixing speed, 58 mg/L of biosolid was recovered at 150 rpm, whereas only about 34 and 39 mg/L were recovered at 100 and 200 rpm, respectively. An optimal mixing speed is required to distribute the bioflocculant throughout the sewage wastewater, allowing the interaction with particles. Lower mixing speeds reduce the distribution of bioflocculant in the wastewater, hindering floc formation. Conversely, higher mixing speeds can break the flocs before they can be recovered [16]. **Figure 5** presents the images of biosolid morphology captured at x10 magnification. The image clearly shows the biosolids flocked together by a bioflocculant.

11. Future research implementations

Knowing the potential of biosolid recovery from sewage wastewater using this treatment, there is a high potential of incorporating the presented results into currently existing sewage treatment plant. The authors suggested the incorporation of a conditioning unit before the sedimentation process [17]. The conditioning unit acted to enhance the settleability of the sewage sludge after the treatment by adding bioflocculant and mixing it under the optimum condition. The resultant sludge from this treatment can be used directly as soil conditioner for agricultural purposes [18]. Depending on the future utilization, future sludge stabilization can also be conducted via anaerobic digestion [19] and/or composting [20].

Before it can be commenced, further optimization and testing on pilot scale need to be conducted to ensure the overall efficiencies of the proposed treatment [21]. Optimizations need to be conducted in terms of the production of bioflocculant, including its mass production cost. In this research, bifoflocculant was used in the form of cell-free supernatant, while solid form is mostly preferred to ease the storage and application [22]. Overall bioflocculant production cost must also include this component. After obtaining the production cost, further cost-benefit and techno-economic analysis should be conducted to assess the feasibility of applying this proposed method to real sewage treatment plant, especially for bigger-scale treatment.

12. Conclusion

Sewage wastewater is produced daily in large quantities around the globe. Proper management of this wastewater can create high-value by-products such as recovered water, nutrients, and biosolids. In addition to by-product recovery, treating this wastewater is essential before it can be released into the environment. Flocculation using environmentally friendly bioflocculant, considering factors such as bioflocculant dosage, type of cation addition, and mixing speed, is an alternative method for treating wastewater and recovering by-products. Increasing the bioflocculant to the optimum dosage enhances collisions and adsorption of organic particles, resulting in high treatment efficiency and biosolid recovery. Different types of cation additions result in varying performance for COD, NH3-N, and biosolid recovery. Furthermore, optimal mixing can distribute the bioflocculant throughout the sewage wastewater, allowing effective interaction with particles for flocculation to occur. This study could advance sewage wastewater treatment processes and enhance society's economy through agricultural activity using nutrient-rich biosolids.

Acknowledgements

This research was financially supported by the Universiti Kebangsaan Malaysia under Geran Universiti Penyelidikan with a grant no. GUP-2024-021.

Author details

Hassimi Abu Hasan[1,2,3]*, Hema Darshinni Vijayakumar[1], Jahira Alias[1], Nur 'Izzati Ismail[1,2,3] and Setyo Budi Kurniawan[1]

1 Department of Chemical and Process Engineering, Faculty of Engineering and Built Environment, Universiti Kebangsaan Malaysia, Selangor, Malaysia

2 Research Centre for Sustainable Process Technology (CESPRO), Faculty of Engineering and Built Environment, Universiti Kebangsaan Malaysia, Selangor, Malaysia

3 Water and Wastewater Treatment Technology Research Group, Universiti Kebangsaan Malaysia, Selangor, Malaysia

*Address all correspondence to: hassimi@ukm.edu.my

References

[1] Awasthi MK, Ganeshan P, Gohil N, Kumar V, Singh V, Rajendran K, et al. Advanced approaches for resource recovery from wastewater and activated sludge: A review. Bioresource Technology. 2023;**384**:1-11. DOI: 10.1016/j.biortech.2023.129250

[2] Nand S, Mishra I, Neeraj A, Naseem M, Patel A, Srivastava PK, et al. Novel plant waste-based cost-effective adsorbent to remove contaminants from sewage wastewater. Groundwater for Sustainable Development. 2024;**24**:1-14. DOI: 10.1016/j.gsd.2023.101072

[3] Sikosana ML, Sikhwivhilu K, Moutloali R, Madyira DM. Municipal wastewater treatment technologies: A review. Procedia Manufacturing. 2019;**35**:1-7. DOI: 10.1016/j.promfg.2019.06.051

[4] Alias J, Hasan HA, Abdullah SRS, Othman AR. Properties of bioflocculant-producing bacteria for high flocculating activity efficiency. Environmental Technology and Innovation. 2022;**27**:1-10. DOI: 10.1016/j.eti.2022.102529

[5] Artifon W, Immich APS, da Silva A, de Souza AAU, de Oliveira D. A novel bioflocculant extracted from excess-activated sludge for dye-containing wastewater treatment. Journal of Water Process Engineering. 2023;**56**:1-8. DOI: 10.1016/j.jwpe.2023.104568

[6] Ogbonna CN, Nwoba EG. Bio-based flocculants for sustainable harvesting of microalgae for biofuel production. A review. Renewable and Sustainable Energy Reviews. 2021;**139**:1-16. DOI: 10.1016/j.rser.2020.110690

[7] Alnawajha MM, Kurniawan SB, Imron MF, Abdullah SRS, Hasan HA, Othman AR. Plant-based coagulants/flocculants: Characteristics, mechanisms, and possible utilization in treating aquaculture effluent and benefiting from the recovered nutrients. Environmental Science and Pollution Research. 2022;**29**:58430-58453. DOI: 10.1007/s11356-022-21631-x

[8] Kurniawan SB, Abdullah SRS, Imron MF, Said NSM, Ismail N'I, Hasan HA, et al. Challenges and opportunities of biocoagulant/bioflocculant application for drinking water and wastewater treatment and its potential for sludge recovery. International Journal of Environmental Research and Public Health. 2020;**17**:1-33. DOI: 10.3390/ijerph17249312

[9] Awasthi MK, Singh E, Binod P, Sindhu R, Sarsaiya S, Kumar A, et al. Biotechnological strategies for bio-transforming biosolid into resources toward circular bio-economy: A review. Renewable and Sustainable Energy. Reviews. 2022;**156**:1-20. DOI: 10.1016/j.rser.2021.111987

[10] HACH. User Manual: DR3900. Edition 12. 2021

[11] Li C, Wu K, Chen L, Liu Z, Zhao X, Li Y, et al. Advanced treatment of low-pollution and poor biodegradability sewage by combined process. Journal of Cleaner Production. 2023;**414**:137366. DOI: 10.1016/j.jclepro.2023.137366

[12] Fang K, Wang B, Zhang Y, Li H. Optimized production and characterization of cation-independent bioflocculant produced by Klebsiella sp. 59L. Environmental Science and Pollution Research. 2020;**28**(7):7981-7993. DOI: 10.1007/s11356-020-11162-8

[13] Agunbiade MO, Van Heerden E, Pohl CH, Ashafa AT. Flocculating performance of a bioflocculant produced by arthrobacter humicola in sewage waste water treatment. BMC Biotechnology. 2017;**17**(1):1-9. DOI: 10.1186/s12896-017-0375-0

[14] Tang W, Song L, Li D, Qiao J, Zhao T, Zhao H. Production, characterization, and flocculation mechanism of cation independent, pH tolerant, and thermally stable bioflocculant from enterobacter sp. ETH-2. PLoS One. 2014;**9**(12):1-19. DOI: 10.1371/journal.pone.0114591

[15] Pan Y, Shi B, Zhang Y. Research on flocculation property of Bioflocculant PG.a21 Ca. Modern Applied Science. 2009;**3**(6):1-7. DOI: 10.5539/mas.v3n6p106

[16] Ding Y, Zhao J, Wei L, Li W, Chi Y. Effects of mixing conditions on floc properties in magnesium hydroxide continuous coagulation process. Applied Sciences. 2019;**9**(5):973. DOI: 10.3390/app9050973

[17] Wang H-F, Hu H, Wang H-J, Zeng RJ. Impact of dosing order of the coagulant and flocculant on sludge dewatering performance during the conditioning process. Science of the Total Environment. 2018;**643**:1065-1073. DOI: 10.1016/j.scitotenv.2018.06.161

[18] Owodunni AA, Ismail S, Kurniawan SB, Ahmad A, Imron MF, Abdullah SRS. A review on revolutionary technique for phosphate removal in wastewater using green coagulant. Journal of Water Process Engineering. 2023;**52**:103573. DOI: 10.1016/j.jwpe.2023.103573

[19] Lu Y, Xu Y, Chen S, Dong B, Dai X. Effect of nitrite addition on the two-phase anaerobic digestion of waste activated sludge: Optimization of the acidogenic phase and influence mechanisms. Environmental Pollution. 2020;**261**:114085. DOI: 10.1016/j.envpol.2020.114085

[20] Nafez AH, Nikaeen M, Kadkhodaie S, Hatamzadeh M, Moghim S. Sewage sludge composting: Quality assessment for agricultural application. Environmental Monitoring and Assessment. 2015;**187**:1-9. DOI: 10.1007/s10661-015-4940-5

[21] Kurniawan SB, Imron MF, Abdullah SRS, Othman AR, Hasan HA. Coagulation–flocculation of aquaculture effluent using biobased flocculant: From artificial to real wastewater optimization by response surface methodology. Journal of Water Process Engineering. 2023;**53**:103869. DOI: 10.1016/j.jwpe.2023.103869

[22] Ahmad A, Kurniawan SB, Ahmad J, Alias J, Marsidi N, Said NSM, et al. Dosage-based application versus ratio-based approach for metal- and plant-based coagulants in wastewater treatment: Merits, limitations, and applicability. Journal of Cleaner Production. 2022;**334**:130245. DOI: 10.1016/j.jclepro.2021.130245

Chapter 4

AIoT Integrated Autonomous Sewage Management

Gorur Bettaiah Janardhana Swamy

Abstract

The AIoT Integrated Autonomous Sewage Management System is a project that aims to develop a cutting-edge solution for effective management and monitoring of sewage systems. The system will leverage Artificial Intelligence of Things (AIoT) technology to autonomously manage the sewage infrastructure, including monitoring the water flow, water quality, and identifying any blockages or leaks in the system. The system will also incorporate a real-time alerting mechanism to notify relevant stakeholders in case of any emergencies or deviations from the expected operating conditions. The project's ultimate goal is to create a self-regulating, eco-friendly sewage management system that ensures minimal human intervention while maximizing efficiency and cost-effectiveness. The AIoT Integrated Autonomous Sewage Management System has the potential to revolutionize the sewage management industry and pave the way for more sustainable and responsible waste management practices.

Keywords: sensors, autonomous sewage management, artificial intelligence, Internet of Things, Artificial Intelligence of Things

1. Introduction

The management of sewage systems is a critical aspect of modern infrastructure, and it is essential to ensure that the system operates efficiently, safely, and sustainably. The conventional sewage management systems require a considerable amount of manual labour, which exposes workers to various hazards, including poisonous gases, bacteria, and viruses.

To address these challenges, the AIoT Integrated Autonomous Sewage Management System has been developed, which utilizes advanced Artificial Intelligence of Things (AIoT) technology to provide an autonomous, safe, and eco-friendly sewage management system. The system integrates various sensors and monitoring devices, which continuously gather data about the sewage system's operation, water flow, and quality, and identify any blockages or leaks in the system [1]. The AIoT technology used in the system enables real-time data analysis and decision-making, reducing the need for human intervention in the sewage management process [2].

The objective of this project is to create a self-regulating, eco-friendly sewage management system that ensures minimal human intervention while maximizing efficiency and cost-effectiveness. The AIoT Integrated Autonomous Sewage Management is a significant step towards a more sustainable and responsible sewage management process, minimizing the human impact on the environment and improving the safety and health of workers involved in sewage management [3–5].

1.1 AI-enabled IoT

AIoT combines the strengths of both AI and IoT. The Internet of Things with AI (AIoT) enables data analysis and comprehension rather than just data collection. Autonomous decision-making is made possible by the AIoT system's insights and recommendations, which are derived from data analysis. The operation of AIoT includes Data collection, Data analysis, Decision-making, Action and Control, and Continuous Learning [6–8]. Some of the relevant applications of AI-enabled IoT are discussed below.

1.1.1 Voice assistants

Voice assistants are cloud-based voice services that serve as users' personal assistants on a tabletop. They use nearby smart gadgets and third-party apps to accomplish a variety of tasks. With voice commands, they can do a lot of things, like playing music, ordering taxis, booking restaurants, answering questions, turning on and off smart lights, and much more. For example, Alexa, Siri, and Google Assistants [6, 9].

1.1.2 Automated vacuum cleaners

iRobot is a well-known company specializing in the development of automated vacuum cleaners, commonly referred to as robot vacuums [6, 8]. Robot vacuums offer the convenience of autonomous cleaning, making them popular for busy households [6, 8].

1.1.3 Smart devices

These devices transform how we interact with technology in our homes, workplaces, and beyond. Smart devices are enabled with Artificial Intelligence (AI) to enhance functionality, improve user experience, and provide automation and insights that were not possible before. Examples and applications of smart devices are discussed below [8].

1.1.3.1 Smart speakers and displays

Smart speakers and displays, such as Amazon Echo with Alexa that integrates AI for voice-controlled assistant capabilities, are used for managing smart home devices, answering questions, and providing entertainment and Google Nest Hub that features a smart display with Google Assistant for controlling smart devices, viewing photographs, and more [8]. *Smart Thermostats*: Nest Learning Thermostat utilizes AI to learn schedules and preferences, optimizing heating and cooling, and Ecobee Smart Thermostat that includes a built-in voice assistant and room sensors for more efficient climate control [8]. *Smart Security Systems*: Ring Video Doorbell features AI-powered motion detection and video surveillance and Arlo Pro that provides wireless security cameras with AI for motion detection and smart notifications [8]. Some of the other examples are Tesla Autopilot [10], BMW iDrive [10], and Intelligent Traffic Management.

1.1.4 Industrial IoT

IoT offers a wide range of applications across numerous industrial sectors, in addition to being utilized inside smart homes [11]. These systems analyze a company's financial and statistical data overall before employing AI and machine learning (ML) algorithms to generate forecasts [6, 12]. The few example solutions are *Primer and Plutoshift*. Primer by Alluvium is designed to enhance operational efficiency through real-time data analysis and anomaly detection. It provides stability scores to monitor equipment health, collecting data from various sensors. Primer uses advanced algorithms for early anomaly detection and predictive maintenance, reducing downtime and costs. Its user-friendly interface ensures easy monitoring and timely alerts, while scalability supports deployment across multiple sites. By offering data-driven insights, Primer improves decision-making, safety, and overall operational efficiency in industrial environments [6, 12]. Plutoshift is an AI-powered platform designed to optimize operational performance, reduce costs, and enhance sustainability across various industries. It leverages machine learning to provide advanced analytics and a unified view of operational data, enabling proactive decision-making and efficient resource management [6, 13]. Therefore, combining AI with IoT can enhance their potential and opportunities. ML and Big Data Analytics (BDA) have the ability to extract extremely valuable insights from the data that IoT creates. The data generated by the Internet of Things are meaningless without AI. Since it is hard for a human to identify information in the data that IoT generates, IoT must rely on AI. Furthermore, the machine will be able to learn on its own, in the event that a new pattern in the data is found, something that a non-AI IoT system will not be able to accomplish [6, 12].

2. Proposed system

The proposed system involves AIoT technology that enables real-time data analysis and decision-making, reducing the need for human intervention in the sewage management process. The system can detect and predict potential issues in the sewage system, such as blockages or leaks, and take appropriate actions to prevent further damage. For example, if the system detects a blockage in the sewage system, it can automatically redirect the flow of sewage to an alternative path, preventing overflow or other hazards.

2.1 Architecture of proposed system

(See **Figure 1**).

2.1.1 System architecture diagram

This diagram could illustrate the various components of the system, including the AIoT sensors, monitoring devices, alerting mechanism, and other system components.

2.1.2 AIoT sensor network diagram

This diagram could illustrate the deployment of AIoT sensors throughout the sewage system, along with their connectivity and data transmission.

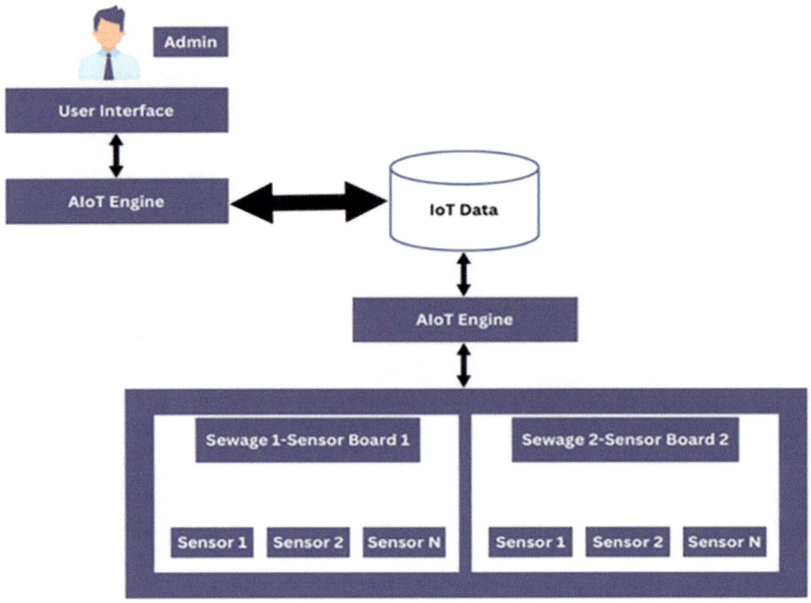

Figure 1.
AIoT Sewage Management System.

The AIoT Integrated Autonomous Sewage Management is an innovative solution to address the challenges associated with traditional sewage management systems. The system utilizes advanced Artificial Intelligence of Things (AIoT) technology to provide an autonomous, safe, and eco-friendly sewage management process.

The system includes various components, such as AIoT sensors, monitoring devices, alerting mechanism, and other system components. The AIoT sensors are deployed throughout the sewage system to gather real-time data about the system's operation, water flow, and quality. These sensors are connected to a central monitoring device that continuously collects, analyses, and processes the data to provide insights into the sewage system's performance.

2.2 Proposed methodology

The proposed methodology for the design and development of AIoT Integrated Sewage Management system has following steps:

Step 1: Sensor Deployment and Data Collection.

Step 2: AIoT Integration.

Step 3: Autonomous Decision-Making.

Step 4: Testing and Validation.

Step 5: Eco-Friendly Optimization.

Step 6: Cost-Effectiveness Analysis.

The methodology for designing the AIoT Integrated Sewage Management system consists of six steps, with each step playing an important function. All steps are explained here.

1. *Sensor Deployment and Data Collection*:

    a. Install water flow sensors, water quality sensors (pH, turbidity, etc.), and leak detectors.

    b. Collect real-time data on water parameters and system status.

2. *AIoT Integration*:

    a. Develop AI models for predictive maintenance, anomaly detection, and blockage identification.

    b. Integrate AI algorithms with sensor data to make informed decisions.

3. *Autonomous Decision-Making*:

    a. Implement decision logic to autonomously manage sewage infrastructure:

        1. Adjust flow rates.

        2. Detect and clear blockages.

        3. Optimize treatment processes.

4. *Testing and Validation*

    a. Simulate various scenarios to validate system performance.

    b. Ensure responsiveness, accuracy, and reliability.

5. *Eco-Friendly Optimization*:

    a. Minimize energy consumption by optimizing pump schedules.

    b. Reduce water wastage through leak detection and timely repairs

6. *Cost-Effectiveness Analysis*:

    a. Evaluate the system's cost savings compared to traditional manual management.

    b. Consider long-term benefits and operational efficiency

2.3 System design

The components, which are used in the implementation of proposed system, are shown below.

2.3.1 AIoT solution

Figure 2 depicts a complete AIoT solution that provides a framework that integrates AI with hardware and IoT software [14, 15].

2.3.2 Gas sensor

Figure 3 shows a gas sensor module that consists of a steel exoskeleton under which a sensing element is housed. Gas sensors used detect the presence and concentration of various gases, providing real-time data that can be used to ensure safety and improve air quality [16–19].

2.3.3 DHT11

Figure 4 depicts DHT11. The DHT11 is a basic, ultra-low-cost digital temperature and humidity sensor. It uses a capacitive humidity sensor and a thermistor to measure the surrounding air and spits out a digital signal on the data pin. Humidity sensors measure and report both moisture and air temperature [16, 20].

2.3.4 Chemical sensor

Figure 5 shows chemical sensors that detect and measure chemical properties in an analyte and convert this information into an electronic signal. These sensors are

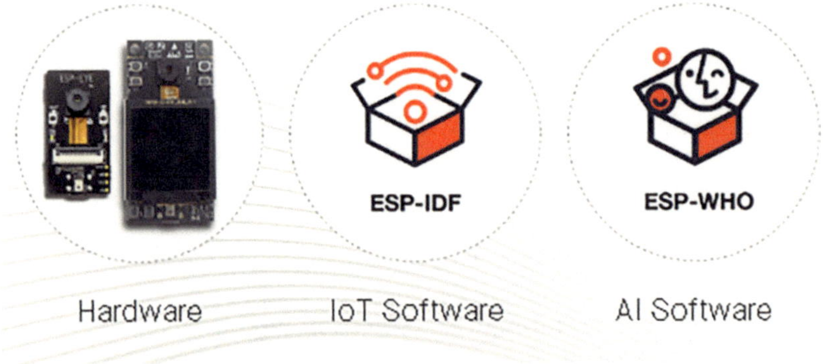

Figure 2.
AIoT Solution.

Figure 3.
Gas Sensor.

Figure 4.
DHT11.

Figure 5.
Chemical Sensor.

widely used in various applications, including environmental monitoring, industrial process control, medical diagnostics, and safety systems [16, 21].

2.3.5 Buzzer

Figure 6 depicts a buzzer, which is an audio signaling device that emits a buzzing or beeping sound. It's commonly used in various applications to provide alerts, notifications, or confirmations. Buzzers can be mechanical, electromechanical, or piezoelectric [18, 22].

2.3.6 LCD

Figure 7 depicts a liquid crystal display (LCD). It is a flat-panel display or another electronically modulated optical device that uses the light-modulating properties of

Figure 6.
Buzzer.

Figure 7.
LCD.

liquid crystals [18]. Liquid crystals do not emit light directly, instead use a backlight or reflector to produce images in color monochrome [18]. LCDs are available to display arbitrary images (as in a general-purpose computer display) or fixed images with low information content, which can be displayed or hidden, such as pre-set words, digits, and seven-segment displays, as in a digital clock [18].

3. Waste water parameters' quality that can be detected by the AIoT system

AIoT systems (Artificial Intelligence of Things) can effectively detect and monitor various wastewater quality parameters. They are discussed as follows.

pH Level: AIoT sensors can continuously measure the acidity or alkalinity of wastewater. Maintaining the right pH is crucial for efficient treatment processes [23, 24]. Temperature: Monitoring water temperature helps optimize biological processes and ensure system stability. Turbidity: Turbidity sensors detect suspended particles in water. High turbidity can indicate pollution or inefficient treatment [23, 24].

Dissolved oxygen (DO): DO levels affect microbial activity. AIoT systems ensure adequate oxygen for biological treatment [23, 24]. Chemical oxygen demand (COD): AI models predict COD levels, aiding in process optimization and resource allocation [23, 24]. Ammonia (NH_3^-): Detecting ammonia helps prevent toxicity and assess nutrient removal efficiency [23, 24]. Nitrate (NO_2^-) and nitrite (NO_1^-): Monitoring nitrogen compounds guides denitrification processes [23, 24]. Total suspended solids (TSS): AIoT systems track solid particles, influencing sedimentation and filtration [23, 24].

Electrical conductivity (EC): It reflects ion concentration and salinity, affecting treatment efficiency [23, 24]. BOD (biochemical oxygen demand): AI predicts BOD levels, crucial for assessing organic pollution [23, 24].

4. Case study on AIoT-based wastewater treatment model for industry 4.0

The case study addresses the AIoT-based treatment of wastewater generated by pharmaceutical companies, which release significant toxins that harm the environment and public health due to high levels of organic and inorganic pollutants. Effective treatment before disposal into the ecosystem is essential. The primary objective is to use industrial data to enhance a wastewater treatment model. Artificial neural network (ANN) algorithms are applied to predict parameters for wastewater plants, enabling users to implement corrective actions and operate the process according to standards. The goal is to achieve improved prediction accuracy in the wastewater treatment model. This study demonstrates the relevance of ANN approaches for predicting input and effluent chemical oxygen demand (COD) in effluent treatment processes. ANNs provide precise technique modeling for complex systems using artificial intelligence methods [25].

4.1 Types of industrial wastewater treatment

Under industrial wastewater processing, the methods and processes used to treat wastewater produced as a by-product of industrial or industrial activity are discussed [25].

4.1.1 Effluent treatment plants (ETPs)

In the chemical and pharmaceutical industries, major firms utilize water purification technologies to remove harmful and non-toxic compounds. Effluent treatment plants (ETPs) play a crucial role in safeguarding the environment by treating wastewater and industrial effluents. During pharmaceutical manufacturing process, pollutants, dust, debris, polymers, and residues are generated and managed through these treatment plants. ETPs employ drying and evaporation processes to handle wastewater. The treatment process aims to eliminate pollutants and minimize contamination risks. A proper wastewater treatment is essential to prevent the build-up of biodegradable organic substances, which could lead to increased pollution, if not addressed promptly. ETPs are strategically arranged to ensure effective treatment and protect environmental health [25, 26].

4.1.2 Sewage treatment plants (STPs)

Domestic wastewater treatment is a method for eliminating impurities using chemical, physical, and biological processes. This treatment removes natural and physiological contaminants, producing a waste stream suitable for environmental reuse. Pre-treatment procedures are crucial for removing raw wastewater materials. During the process, sewage water is treated to eliminate various impurities, resulting in clean water. This treated water can be reused for household or commercial purposes, contributing to water conservation and reducing environmental impact. The outcome is a sustainable solution for managing wastewater and protecting public health and ecosystems [25].

4.1.3 Common and combined effluent treatment plants (CETPs)

Healing systems are unsuitable for small industries, making CETPs a viable alternative. Located near small industrial units, CETPs aim to reduce the costs associated with effluent treatment. These common and integrated systems help small businesses process wastewater efficiently and economically, offering a cost-effective solution for managing industrial effluents [25].

4.2 Wastewater treatment techniques

There are some key techniques for wastewater treatment:

1. Physical treatment: Includes processes like screening, sedimentation, and filtration to remove large particles and suspended solids [26].

2. Biological treatment: Utilizes microorganisms to decompose organic matter. Common methods include activated sludge and trickling filters [26].

3. Chemical treatment: Involves the use of chemicals to remove contaminants. Techniques include coagulation, flocculation, and disinfection [26].

4. Advanced treatment: Employs methods like membrane filtration, reverse osmosis, and advanced oxidation to remove specific pollutants and achieve higher purity levels [26].

4.3 ANN model development steps

Figure 8 illustrates the ANN modeling technique, comprising multiple steps: training data collection, pre-processing the data collected, selecting the ANN structure, ANN parameters determination, the training of ANN, and training failures analysis [25].

4.4 Summary

Artificial neural networks (ANNs) offer a promising approach for predicting and forecasting water variables. This case study demonstrates that ANN-based COD predictions outperform traditional mathematical modeling. Wastewater treatment using ETP involves complex, non-linear processes across physical, chemical, and

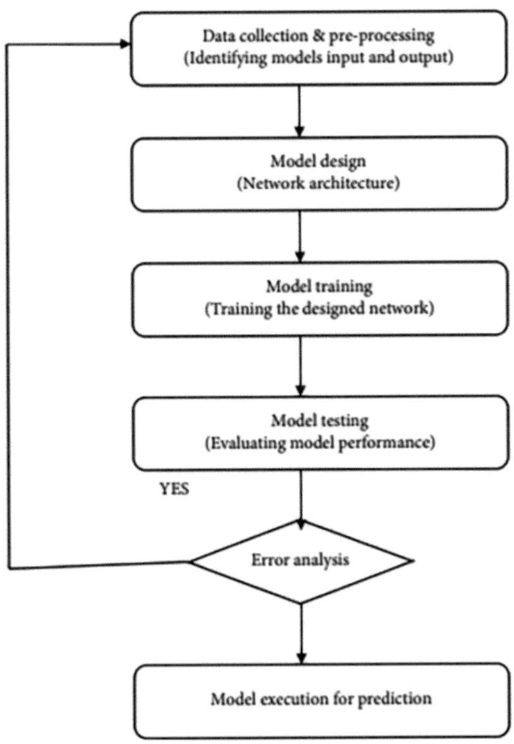

Figure 8.
Steps of ANN Model Development.

biochemical dynamics. Despite these challenges, ANN consistently delivers highly effective results [25, 26].

With advancements in technology, ANN utilizes past plant data to achieve more accurate results. Future work may involve extending the model to optimize wastewater treatment performance in various analyzed states. Employing swarm intelligence techniques is anticipated to enhance outcomes further [25, 26].

5. Challenges of implementing AIoT-enabled autonomous sewage management system

Implementing an AIoT-enabled autonomous sewage management system presents several challenges, including technical, operational, and financial aspects. Here are the key challenges:

5.1 Technical challenges

5.1.1 Integration with legacy systems

a. *Compatibility issues*: Integrating new AIoT technology with existing infrastructure can be complex. Legacy systems may not support modern sensors or data protocols, requiring significant modifications or upgrades.

b. *Data integration*: Consolidating data from various sources into a unified platform for analysis and decision-making can be technically challenging.

5.1.2 Data management

a. *Volume of data*: AIoT systems generate vast amounts of data that need to be processed and analyzed in real-time. Managing and storing these data can strain the existing IT infrastructure.

b. *Data accuracy*: Ensuring the accuracy and reliability of data collected from sensors is critical for effective AI analysis and decision-making.

5.1.3 Security and privacy

a. *Cybersecurity risks*: IoT devices are vulnerable to cyber-attacks. Securing these devices and the data they collect are crucial to prevent unauthorized access and data breaches.

b. *Data privacy*: Handling sensitive data, such as information about wastewater composition and infrastructure status, requires strict privacy measures.

5.2 Operational challenges

5.2.1 Maintenance and reliability

a. *System maintenance*: AIoT systems require regular maintenance to ensure that sensors and other hardware components remain functional. Failure of these components can disrupt system operations.

b. *Reliability*: Autonomous systems must be highly reliable to avoid failures that could lead to environmental hazards or operational issues.

5.2.2 Skill requirements

a. *Technical expertise*: Implementing and managing AIoT systems requires specialized knowledge in AI, IoT, and data analytics. Training staff or hiring experts can be challenging and costly.

b. *Change management*: Transitioning to an AIoT-enabled system may require significant changes in workflow and management practices, which can be difficult to implement.

5.2.3 System calibration

a. *Calibration needs*: AI algorithms need to be calibrated and fine-tuned to accurately predict and manage sewage system issues. This process can be time-consuming and requires ongoing adjustments.

5.3 Financial challenges

5.3.1 Initial costs

a. *Capital investment*: The upfront costs for purchasing and installing IoT sensors, AI software, and supporting infrastructure can be substantial.

b. *Cost-benefit justification*: Demonstrating the return on investment (ROI) for such systems can be challenging, especially when compared to traditional methods.

5.3.2 Operational costs

a. *Ongoing expenses*: Maintaining and operating an AIoT system involves ongoing costs for software updates, data storage, cybersecurity measures, and system maintenance.

b. *Unexpected costs*: Unforeseen technical issues or failures can lead to additional costs for repairs and system adjustments.

5.4 Regulatory and compliance issues

5.4.1 Regulatory compliance

a. *Standards and regulations*: Complying with industry standards and regulations related to wastewater management and data privacy can be complex.

b. *Certification*: Ensuring that AIoT systems meet regulatory requirements and obtain necessary certifications may involve additional time and expense.

5.5 Environmental and social challenges

5.5.1 Public perception

a. *Acceptance*: Gaining public trust and acceptance for AIoT systems, especially regarding data privacy and the reliability of autonomous operations, can be challenging.

b. *Transparency*: Ensuring transparency in how AIoT systems operate and how data are used can help address public concerns.

5.5.2 Environmental impact

Unintended consequences: There is a risk that autonomous systems may inadvertently cause environmental harm if they are not properly calibrated or if they fail to function as intended.

6. Conclusion

The AIoT Integrated Autonomous Sewage Management and Alerting System is an innovative solution to address the challenges associated with traditional sewage management systems. The system utilizes advanced Artificial Intelligence of Things (AIoT) technology to provide an autonomous, safe, and eco-friendly sewage management process. The system integrates various sensors and monitoring devices that continuously gather data about the sewage system's operation, water flow, and quality, and identify any blockages or leaks in the system. The AIoT technology used in the system enables real-time data analysis and decision-making, reducing the need for human intervention in the sewage management process.

The AIoT Integrated Autonomous Sewage Management System is a significant step towards a more sustainable and responsible sewage management process, minimizing the human impact on the environment and improving the safety and health of workers involved in sewage management.

Author details

Gorur Bettaiah Janardhana Swamy
Department of CSE, The National Institute of Engineering, Mysore, India

*Address all correspondence to: jdsmem@gmail.com

References

[1] Iqbal U, Bin Riaz MZ, Barthelemy J, Perez P. Artificial intelligence of things (AIoT)-oriented framework for blockage assessment at cross-drainage hydraulic structures. Australasian Journal of Water Resources. 2023;**27**:1-11

[2] Ahmed M, Qureshi KN. Tri-tier architectures for AIoT networks. In: Artificial Intelligence of Things (AIoT). Boca Raton, Florida, USA: CRC Press; 2024. pp. 34-48

[3] Sung WT, Devi IV, Hsiao SJ, Fadillah FN. Smart garbage bin based on AIoT. Intelligent Automation & Soft Computing. 2022;**32**(3):1387-1401

[4] Joshi HC, Das S. Design and simulation of smart helmet for coal miners using ZigBee technology. International Journal on Emerging Technologies. 2017;**8**(1):196-200

[5] Sharma R. Leveraging AI and IoT for sustainable waste management. In: Whig P, Silva N, Elngar AA, Aneja N, Sharma P, editors. Sustainable Development through Machine Learning, AI and IoT, ICSD 2023. Communications in Computer and Information Science. Vol. 1939. Cham: Springer; 2023

[6] Ghosh A, Chakraborty D, Law A. Artificial intelligence in internet of things. CAAI Transactions on Intelligence Technology. 2018;**3**(4):208-218

[7] Merenda M, Porcaro C, Iero D. Edge machine learning for AI-enabled IoT devices: A review. Sensors. 2020;**20**(9):2533

[8] Gubbi J, Buyya R, Marusic S, Palaniswami M. Internet of things (IoT): A vision, architectural elements, and future directions. Future Generation Computer Systems. 2013;**29**(7):1645-1660

[9] Skorikov M, Omar KNJ, Khan R. Voice-controlled intelligent personal assistant. In: García Márquez FP, editor. International Conference on Intelligent Emerging Methods of Artificial Intelligence & Cloud Computing, IEMAICLOUD 2021. Smart Innovation, Systems and Technologies. Vol. 273. Cham: Springer; 2022

[10] Moorhouse GV. Natural User Experience in Tertiary Driver-Car Interactions. Italy: Politecnico di Milano; 2022

[11] Sehrawat D, Gill NS. Smart sensors: Analysis of different types of IoT sensors. In: 2019 3rd International Conference on Trends in Electronics and Informatics (ICOEI). Piscataway, New Jersey, USA: IEEE; Apr 2019. pp. 523-528

[12] Bhushan B, Sangaiah AK, Nguyen TN. AI Models for Blockchain-Based Intelligent Networks in IoT Systems. Berlin, Germany: Springer Nature; 2023

[13] Plutoshift. About [Internet]. Plutoshift. Available from: https://www.plutoshift.com/about/ [Accessed: September 12, 2024]

[14] Espressif Systems. ESP-EYE Development Kit [Internet]. Espressif. Available from: https://www.espressif.com/en/products/devkits/esp-eye/overview [Accessed: September 12, 2024]

[15] 99Tech. ESP32 ESP-EYE [Internet]. 99Tech. Available from: https://99tech.com.au/product/esp32-esp-eye-source [Accessed: September 12, 2024]

[16] Janardhana Swamy GB, Janardhana DR, Vijay CP, Ravi V. Blockchain-enabled IoT integrated autonomous sewage management system. In: Convergence of Internet of Things and Blockchain Technologies. Cham: Springer International Publishing; 2021. pp. 41-56

[17] Sereni JGR. Reference Module in Materials Science and Materials Engineering. Amsterdam, Netherlands: Elsevier; 2016

[18] Kuria KP, Robinson OO, Gabriel MM. Monitoring temperature and humidity using arduino nano and module-DHT11 sensor with real time DS3231 data logger and LCD display. International Journal of Engineering Research & Technology (IJERT). Dec 2020;**9**(12):416-422. ISSN: 2278-0181

[19] Abishek M, Divya B, Jeevanantham S, Surendar G, Jagadeeshraja M. Estimation and prediction of Lpg gas using data analytics. International Journal of Engineering Research & Technology (IJERT) Etedm. 2022;**10**(8):211-215. ISSN: 2278-0181

[20] Abu Gharbyeh E. IOT–Based Weather Reporting System. Piscataway, New Jersey, USA: IEEE; 2022

[21] Wang W. Introductory Chapter: What Is Chemical Sensor? London, United Kingdom: InTech; 2016. DOI: 10.5772/64626

[22] Elprocus. Buzzer: Working and Applications [Internet]. Elprocus. Available from: https://www.elprocus.com/buzzer-working-applications/ [Accessed: September 12, 2024]

[23] Jan F, Min-Allah N, Düştegör D. IoT based smart water quality monitoring: Recent techniques, trends and challenges for domestic applications. Water. 2021;**13**(13):1729

[24] Hassan Z, Hossain GJ, Islam MM. Internet of things (IoT) based water quality monitoring system. Educational Research. 2020;**2**(4):168-180

[25] Narendar Singh D, Murugamani C, Kshirsagar PR, Tirth V, Islam S, Qaiyum S, et al. IOT based smart wastewater treatment model for industry 4.0 using artificial intelligence. Scientific Programming. 2022;**22**(1):5134013. DOI: 10.1155/2022/5134013

[26] Nagda A, Meena M, Shah MP. Bioremediation of industrial effluents: A synergistic approach. Journal of Basic Microbiology. 2021;**62**(3-4):395-414. DOI: 10.1002/jobm.202100225

Chapter 5

Socio-Economic and Environmental Effects of Difference Scale in Sewage Sludge Recycling System Considering Ecological Value in China

Jiawen Zhang and Toru Matsumoto

Abstract

With China's ongoing economic development and increasing emphasis on environmental protection, the number of sewage plants is increasing annually and sludge production is increasing. This study analyzed the scale effect of sludge treatment and recycling systems (STRS) with different technologies (i.e., incineration, aerobic composting, used in material, anaerobic digestion) on the performance of environment an socio-economy by life cycle assessment (LCA) and life cycle cost (LCC). The LCA results showed that aerobic composting had a better impact on climate change (−1.53E−03 kg CO_2 eq/p/yr) than other methods, while the whole life cost (WLC) results showed the scenario of using in material had the less cost in four scenarios. Moreover, the socio-economic impact of introducing a carbon ecological compensation mechanism into the STRS to achieve carbon neutrality was analyzed taking Tianjin as an example. In the future, it is recommended to consider and incorporate the environmentally friendly impacts of STRS with various scales into the carbon ecological compensation mechanism.

Keywords: sludge recycle, urban ecological system, ecological value, cost-benefit analysis, carbon neutrality

1. Introduction

The acceleration of economic development and urbanization in China had resulted in a significant increase in the production of dry sludge, reaching 13.69 million tons by 2022, as illustrated in **Figure 1**. Over the past five years, the disposal ratio had been above 95% on average [1]. The issue of sludge disposal had emerged as a significant challenge in the context of wastewater treatment plants and municipal waste management. In contrast, the philosophy of the "circular economy" viewed waste-activated sludge (WAS) as a renewable resource with high levels of organic matter and nutrients [2]. The STRS system could generate a variety of energy forms

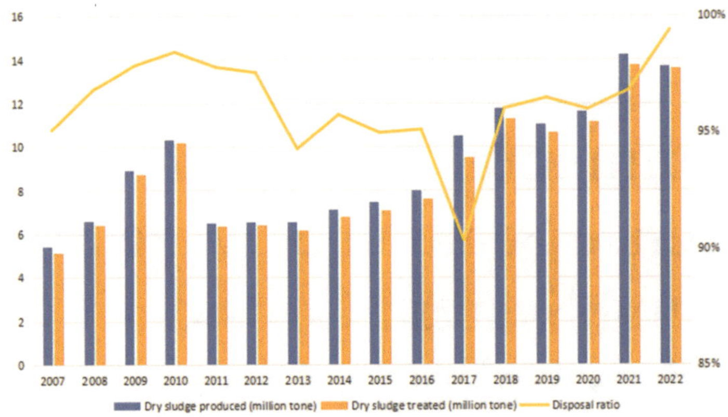

Figure 1.
The situation of sewage sludge produced and treated proportion in China.

and products, which had the potential to avoid greenhouse gas (GHG) emissions [3, 4]. Examples of these included electricity, biogas [5, 6], fertilizer [7, 8], and building materials [9, 10]. It was evident that the current treatment and disposal management of waste activated sludge (WAS) in China lacked a comprehensive and integrated plan that simultaneously considers environmental protection, economic viability, and social acceptance. Consequently, the establishment of a reliable and holistic assessment methodology that encompasses the overall performance of the STRS system was of paramount importance. This would ensure the optimal utilization of WAS, thus reducing its environmental impact, while also ensuring the financial viability of the system and the social acceptability of its operations.

In September 2020, China announced at the United Nations General Assembly that it would strive to achieve "carbon peak" by 2030 and "carbon neutrality" by 2060. It was widely acknowledged that the increase in GHG emissions resulting from human activities was the primary driver of global warming. The sector of various waste treatment, such as household waste and wastewater, was contributed 1.7% to total GHG emission of China in 2020, achieved 209.5 billion tones included wastewater treatment. The wastewater treatment had been identified as a significant source of anthropogenic GHGs in the urban.

Wastewater treatment plants (WTTPs) have the capacity to generate carbon dioxide (CO_2), methane (CH_4), and and nitrous oxide (N_2O) through a variety of chemical and biological processes, as well as energy production and combustion. The GHG emission boundaries of wastewater treatment systems had not been clearly defined. Previous research had overlooked the GHG emissions associated with sludge transportation and treatment, which can contribute up to 40% of the total emissions from WTTPs. Due to the vast size of China and the significant disparities in economic development, technological advancement, and industrial structure among its provinces, there were notable variations in sludge disposal strategies, as illustrated in **Figures 2** and **3**. Consequently, the challenge of reducing carbon emissions in sewage treatment systems varied considerably. It was imperative to develop tailored sludge disposal plans specific to each region.

In the preceding studies, LCA and LCC had been employed extensively in order to evaluate the environmental and economic impacts of various sludge management schemes [11–15]. However, the combined environmental and economic impacts of the WASR system when considering alternative production scenarios had yet to be fully elucidated in the existing literature. To address these limitations and meet the

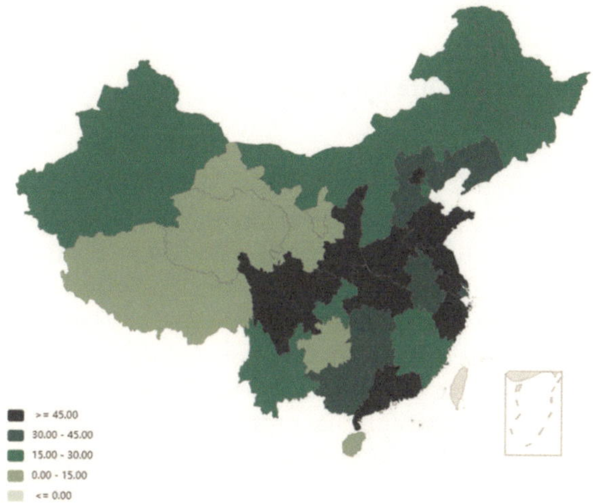

Figure 2.
The dry sludge produced of each province in China.

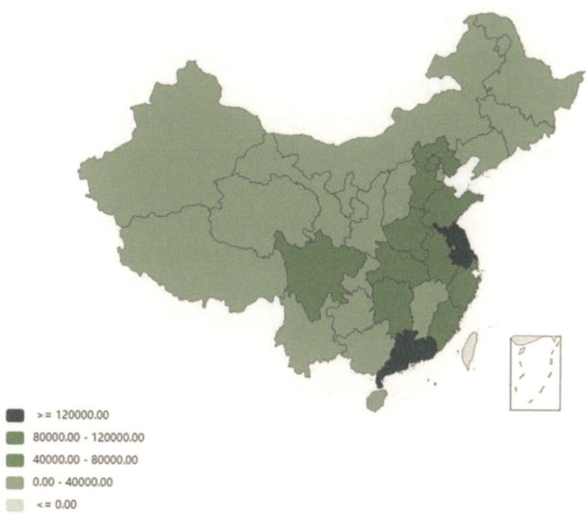

Figure 3.
The gross regional product of each province in China.

practical needs of the industry, we employed whole life costing (WLC) in accordance with the guidelines set forth in BSI ISO 15686-5 (2008). This approach integrated the two aspects by including the external cost, which represented the monetized value of environmental pollutants.

Accordingly, the objective of this study was to integrate the environmental and economic evaluation models of four WASR systems in China *via* LCA and LCC methods and to propose future optimization scenarios. The study also determined the main contribution of pollutants to external costs, which marked the potential social damages caused by pollutant discharge for STRS management. In order to achieve the carbon peak target as soon as possible, and in consideration of the variations in economic development among regions in China, this study calculated the cost of the sludge management scheme to address regional disparities through the carbon ecological compensation mechanism.

2. Methodology

2.1 Life cycle assessment and life cycle cost

2.1.1 Goal and scope definition

The objective of this research was to evaluate the environmental and economic efficiency of four prevalent technologies for treating and recycling WAS in China and to analyze them under various policy scenarios. The study focused on processing 1 ton of WAS with 80% moisture content as the functional unit. The system encompassed transportation, drying, pretreatment, production, and air pollution control measures. The outputs from the STRS, such as electricity, fertilizer, bricks, and biogas, were considered as substitutes for conventional products in the market. System expansion was implemented to prevent the allocation of any by-products during the process (**Figure 4**).

2.1.2 Data source and life cycle inventory

The data inventory for various processes such as incineration, aerobic composting, brick production, and anaerobic digestion were gathered from enterprises *via* environmental impact assessment (EIA) reports. The life cycle inventories of the original main products were derived from the Chinese Life Cycle Database by software Ebalance and previous literature [16–19]. Factors such as socioeconomic conditions, technology efficiency, and the quality of WAS could significantly impact both environmental and economic performance. Therefore, a case-by-case evaluation of each WAS from wastewater treatment plants was necessary [14]. To facilitate a comparative analysis, it was assumed that the quality of WAS adheres to the standard GB 24188-2009, and regional differences did not substantially influence the evaluation outcomes in this study.

Economic data regarding budget costs were sourced from market research and EIA reports of individual enterprises. Operational costs considered in this study

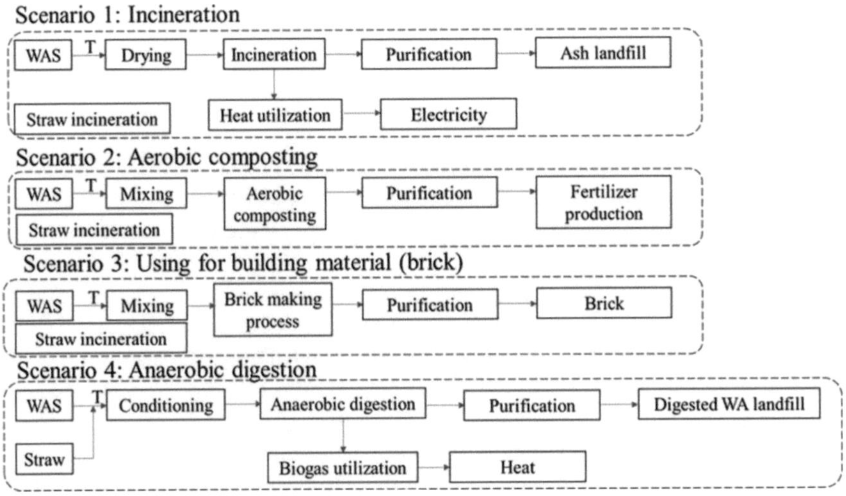

Figure 4.
System boundaries of four scenarios of WAS recycling system.

encompassed expenses related to transportation, raw materials, energy consumption, and labor. Sales incomes were inclusive of all revenues generated from the sale of final products. Prices of final products and raw materials were obtained from publicly available sources, reflecting their market rates in 2019. The lack of a standardized WAS disposal subsidy across Chinese cities led this study to adopt the subsidy amount in Chongqing (205 CNY/t). The assumed distance for transportation between each technology was set at 100 km.

The application of the external cost valuation method from Japan to China was deemed unfeasible due to differing economic development stages and environmental perspectives [20]. Nonetheless, the external costs were determined based on the Environmental Protection Law in China and CREO 2017. In cases where data on external costs were unavailable, the monetization value of per unit total emissions was referenced from prior studies [16].

2.1.3 Environmental and economic evaluation

This research assessed and measured the environmental performance of WASR through a Life Cycle Assessment (LCA) using the ReCipe 2008 [14, 21]. LCA was chosen as the preferred approach due to its comprehensive coverage of potential environmental impacts [22]. The study focused on quantifying the total environmental impact of climate change (EPtotal) resulting from both direct and indirect greenhouse gas emissions during the operation of the WASR system. Additionally, the avoided environmental impact of climate change (EPavoid) was determined, which represented the environmental impact of products replacing an equivalent number of traditional products. The net environmental impact of climate change (EPnet) was calculated to provide a more accurate assessment of the true environmental impact of each scenario.

$$EP_{net} = EP_{total} - EP_{avoid} \tag{1}$$

The calculation of the life cycle cost (LCC) was conducted and evaluated based on the net present value, as Eq. (2). The cash inflow (C_I) represented the total amount of monetary inflows in the initial year, including the WAS subsidy, revenue from final product sales, and income from carbon compensation in each scenario. Conversely, the cash outflow (C_O) encompassed the total amount of monetary outflows in the initial year, comprising capital expenditures, operational expenses, and tax payments [15].

$$LCC = C_I - C_O \tag{2}$$

The WLC analyzed LCC and external cost considerations to evaluate the comparative performance of different scenarios. The study applied the concept of “true cost” as outlined in the BSI ISO 15686-5 standard. The external costs for each scenario were calculated using the Ecotax 2002 method in Sweden, as described in reference [23, 24]. This involved multiplying the quantity of emissions k ($e_{j,k}$) by the unit price of emissions k (P_k) for each scenario. The currency exchange rate utilized in the research was 1 USD = 6.804 RMB.

$$\text{Externality cost}_j = \sum\nolimits_{j=0}^{4} t_j \cdot \sum_{j,k} e_{j,k} \cdot P_k \tag{3}$$

The external costs in each scenario primarily were considered the financial damages caused by water quality, human health, climate change, and indeterminate issues [20, 23, 24]. The external costs related to climate change involve the quantified worth of CO_2 and CH_4.

2.2 The effect of implement scale of sewage sludge recycling system

Previous studies had found that the correlation between the implementation scope, ecological emissions, and financial expenditures of STRS were consistent with the least squares method. By conducting a cost-benefit analysis that took into account the impact of the implementation scale, the most effective system and cost-balanced scale could be determined. Following an in-depth examination, it was found that the correlation between environmental emissions and economic expenditures was closely related to the implementation scope. This relationship could be measured by a power function, as shown in **Table 1** [25].

The environmental performance by climate change of each province was calculated by Eq. (4).

$$\mathrm{EP} = \sum \frac{EP_{i,base}}{\left(\frac{S_{base}}{S}\right)^{x_{GHG}}} \tag{4}$$

In this context, EP referred to the environmental performance of each province in climate change. $\mathrm{EP}_{\mathrm{i,base}}$ referred to the environmental performance of the baseline for scenario i in climate change. S_{base} was the implementation scale of the baseline. S was the implementation scale of the target scenario. x_{GHG} was the scaling exponent of environmental performance for scenario i.

2.3 Carbon ecological compensation mechanism

Carbon ecological compensation, a new area of ecological compensation research, aims to adjust the interest dynamics between ecological protection areas and beneficiary regions by facilitating the flow of funds between them. This approach helps narrow the development gap between regions, alleviate financial pressures on ecological protection areas to some extent, and ultimately achieve ecological justice. It has emerged as a crucial strategy for numerous countries worldwide to address climate change and promote coordinated regional development. In carbon ecological compensation, the central government is responsible for establishing a horizontal carbon ecological compensation mechanism. The primary objective is to promote the coordinated development of the regional economy and ecological environment, as well as to

Scenario	The implement scale of baseline	Unit cost	Unit GHG emission
Incineration	200 t/d	0.438	0.476
Aerobic composting	400 t/d	0.515	0.350
Used in brick	15,000 t/a (60% water content)	0.606	0.595
Anaerobic digestion	50 t/d	0.576	0.479

Table 1.
The implement scale of STRS in economic and GHG emission sector.

incentivize and guide provinces in utilizing sludge resources. The provincial governments aim to optimize the environmental and economic benefits of the sludge treatment system within the region by leveraging existing resources under the horizontal carbon ecological compensation mechanism. Assuming that the forest carbon sink in the region, shown in **Table 2**, serves as the baseline emission, if the carbon emissions from the sludge treatment system exceed the baseline emission, carbon credits must be purchased. Conversely, if the carbon emissions are lower than the baseline, carbon credits will be earned.

3. Result and discussion

In terms of climate change for scenario 1, CO_2 contributed impact of climate change by about 97.51%. The main source is energy consumption, of which low calorific value and high water content are the main influences. The finding of this analysis is in alignment with those of Guo et al. [26]. Compared with aerobic composting, the method of incineration consumed more energy to reduce the water content. Energy consumtion was the main contributor of climate change and operation cost. According to the results of the climate change in **Table 3**, scenario 2 has a favorable effect on the environment. The reduction in N_2O is the primary factor contributing to the favorable environmental impact of scenario 2 in climate change. The process of making mineral fertilizers releases N_2O, which has a significant impact on climate change. Compared to the manufacture of mineral fertilizer, scenario 2 does not result in the generation of N_2O. Analysis of the components of climate change indicates that CO_2 is the primary pollutant responsible for climate change. The primary sources of CO_2 are energy use and the amount of organic matter present. On the

Province	Forest carbon sequestration	Province	Forest carbon sequestration
Beijing	217.74	Henan	853.55
Tianjin	38.60	Hubei	1912.75
Hebei	657.60	Hunan	2045.85
Shanxi	866.41	Guangdong	2847.40
Inner Mongolia	4137.13	Guangxi	4313.57
Liaoning	1139.17	Hainan	1522.92
Jilin	2046.77	Chongqing	1616.11
Heilongjiang	5162.03	Sichuan	4547.38
Shanghai	65.82	Guizhou	2335.63
Jiangsu	266.04	Yunnan	5961.94
Zhejiang	1658.99	Shaanxi	1994.37
Anhui	1027.61	Gansu	1003.75
Fujian	3020.66	Qinghai	155.81
Jiangxi	2440.29	Ningxia	55.28
Shandong	157.50	Xinjiang	1809.94

Table 2.
Forest carbon sequestration by province in China in 2017 (unit: 10^4 tone).

Category	Unit	S1	S2	S3	S4
Climate Change	kg CO_2 eq/p/yr	4.94E−03	−1.53E−03	6.96E−04	7.41E−05

Table 3.
Global warming impact of each scenario for STRS.

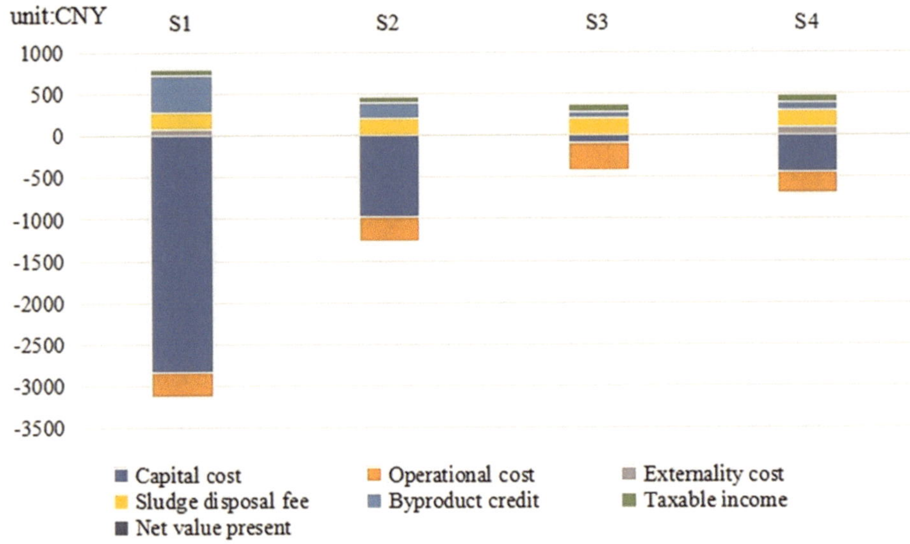

Figure 5.
The WLC of each scenario for STRS.

one hand, the sludge's organic matter level needs to be decreased. Sludge treatment and recycling, on the other hand, ought to prioritize energy conservation. The main influencing constituents of climate change in scenario 4 are CO_2 (94.27%) as same as in scenario 1.

The primary expenses of the system were further examined to offer recommendations for enhancement, as depicted in **Figure 5**. The findings reveal that the primary cost associated with the system is the construction cost as shown in **Figure 6**. Within these expenses, transportation costs constitute over 56% of the operational expenses.

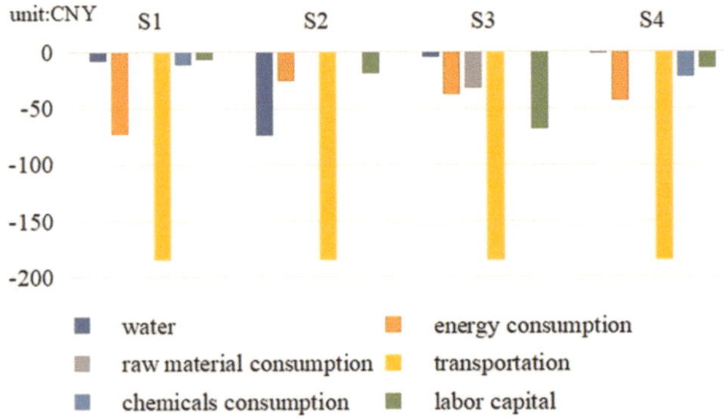

Figure 6.
The optional cost of each scenario for STRS.

Therefore, it is advisable for future management of the WAS system to strategically plan the reuse location and optimize transportation routes. In scenarios 2 and 3, in addition to transportation costs, water expenses play a significant role, with the increasing moisture content of WAS upon exiting the sewage treatment plant necessitating attention. It is crucial to strike a balance between transportation and water costs in future management, as the volume is correlated with the moisture content of WAS, and transportation expenses escalate with the volume of WAS. In alternative scenarios, apart from transportation costs, energy expenditures such as electricity, natural gas, and coal also serve as major cost contributors, and their financial performance can be improved by enhancing energy efficiency.

The government promotes the use of sludge as a resource, which can generate job opportunities for citizens and boost public environmental consciousness by involving them in environmental protection. Large-scale recycling systems have the potential to contribute significantly to job creation in the waste management sector. By employing a larger workforce to manage and operate these systems, job opportunities are created, thereby stimulating economic growth. On the other hand, small-scale systems promote local entrepreneurship and community involvement by encouraging individuals or small businesses to engage in waste recycling activities. These systems not only empower local communities but also foster a sense of environmental responsibility. Additionally, medium-scale systems can enhance regional economic development through resource recovery projects that generate revenue and promote sustainable practices.

Taking Tianjin as an example, there were 12 sludge treatment plants in 2021, including facilities for industrial sludge treatment. The dry sewage sludge produced in Tianjin was 170,212 tons. The environment performance of climate change of total sludge treatment plants in Tianjin was -0.021 tCO_2 eq calculated by Eq. (4) as depicted in **Figure** 7, P1, P4, P10, and P12 exhibited negative environmental impacts. The main reason is that these four treatment plants use aerobic composting technology with a daily processing capacity of more than 300 tons. This technology replaces traditional fertilizers with sludge, significantly reducing greenhouse gas emissions from the entire system.

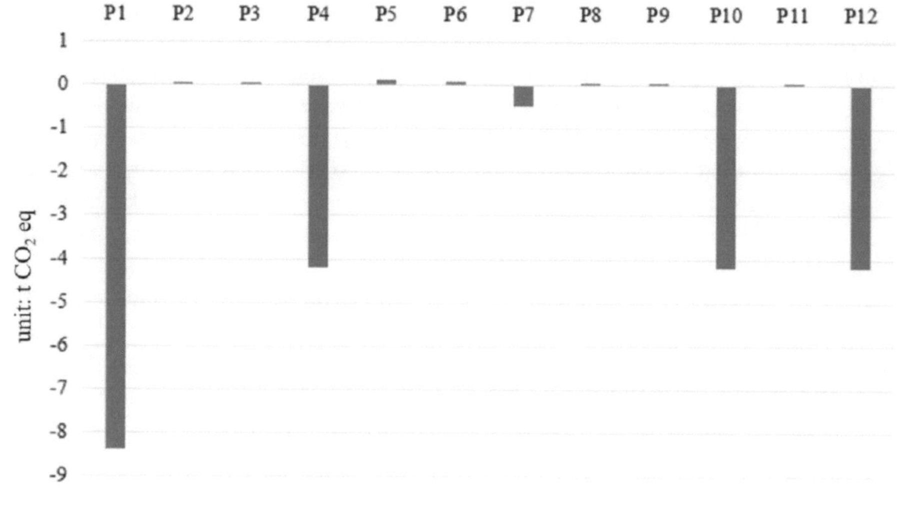

Figure 7.
The effect of climate change of each sludge treatment plants in Tianjin.

This research utilized the mean transaction price of carbon emission rights in the national carbon trading market as the benchmark for carbon compensation, a common practice in contemporary studies on carbon ecological compensation. In 2022, the average annual transaction price in the national carbon trading market stood at a mere 55.3 yuan/t. For instance, in the case of Tianjin's sludge resource utilization, 2135.7 yuan in subsidies was obtained through the reduction of carbon emissions over the course of a year, which significantly undervalues the worth of carbon. The persistent setting of ecological compensation standards at a low level could have adverse effects, such as dampening the drive for ecological development in carbon compensation regions and potentially leading carbon-paying regions to view it as a cost-effective production approach. This could diminish incentives for emission reduction and impede the establishment of effective constraints on carbon-paying regions.

In the future, it is recommended to consider and incorporate the environmentally friendly impacts of STRS with various scales into the carbon ecological compensation mechanism. For example, large-scale systems play a crucial role in decreasing greenhouse gas emissions by effectively handling organic waste and harnessing methane for energy generation. This not only helps combat climate change but also aids in the production of renewable energy. Conversely, small-scale systems reduce pollution related to transportation by treating waste on-site, thereby lessening the carbon footprint linked to waste transportation. Additionally, medium-scale systems support biodiversity and soil health by incorporating ecological principles into their recycling methods, such as utilizing compost as a soil enhancer to enhance soil fertility and structure.

4. Conclusion

Considering the ecological value in sewage sludge recycling systems is essential for promoting sustainability and resilience in waste management practices. Implementing circular economy principles, such as recycling and reusing resources, can maximize resource efficiency and minimize waste generation in these systems. Furthermore, incorporating ecological design elements, such as green infrastructure and nature-based solutions, can enhance ecosystem services and resilience in recycling processes. Establishing partnerships between government, industries, and local communities is crucial for promoting sustainable sewage sludge management practices that prioritize ecological value, ensuring a harmonious balance between socio-economic development and environmental conservation. In conclusion, the scale at which sewage sludge recycling systems operate in China has far-reaching socio-economic, environmental, and ecological implications. By understanding and harnessing the socio-economic benefits of different scales, mitigating environmental impacts, and prioritizing ecological value, China can build a more sustainable and resilient sewage sludge management system that promotes long-term prosperity and environmental health. It is imperative for stakeholders to work together to implement innovative solutions that address the complex challenges posed by sewage sludge management, ensuring a greener and more sustainable future for all.

Conflict of interest

The authors declare no conflict of interest.

Author details

Jiawen Zhang[1*] and Toru Matsumoto[2]

1 Research Institute for Environmental Innovation (Tianjin Binhai), Tianjin, China

2 Institute of Environmental Science and Technology, The University of Kitakyushu, Kitakyushu, Japan

*Address all correspondence to: tjzhangjiawen@163.com

References

[1] China Statistical Yearbook. China: China Statistics Press; 2019. Available from: http://www.stats.gov.cn/tjsj/ndsj

[2] Spinosa L, Doshi P. Re-thinking sludge management within the sustainable development goal 6.2. Journal of Environmental Management. 2021;**287**:112338

[3] Liu BB, Wei Q, Zhang B, Bi J. Life cycle GHG emissions of sewage sludge treatment and disposal options in Tai Lake watershed, China. Science of the Total Environment. 2013;**447**:361-369

[4] Rizzardini C, Goi D. Sustainability of domestic sewage sludge disposal. Sustainability. 2014;**6**(Pt. 5):2424-2434

[5] Liu Y, Lin R, Ren J. Developing a life cycle composite footprint index for sustainability prioritization of sludge-to-energy alternatives. Journal of Cleaner Production. 2021;**281**:124885

[6] Rulkens W. Sewage sludge as a biomass resource for the production of energy: Overview and assessment of the various options. Energy and Fuels. 2008; **22**(Pt. 1):9-15

[7] Elmi A, Al-Khaldy AA, AlOlayan M. Sewage sludge land application: Balancing act between agronomic benefits and environmental concerns. Journal of Cleaner Production. 2020;**250**: 119512

[8] Chu L, He W. Toxic metals in soil due to the land application of sewage sludge in China: Spatiotemporal variations and influencing factors. Science of the Total Environment. 2021;**757**:143813

[9] Limami H, Manssouri I, Cherkaoui K, Khaldoun A. Recycled wastewater treatment plant sludge as a construction material additive to ecological lightweight earth bricks. Cleaner Engineering and Technology. 2021;**2**: 100050

[10] Okuno N, Ishikawa Y, Shimizu A, Yoshida M. Utilization of sludge in building material. Water Science and Technology. 2004;**49**(Pt. 10):225-232

[11] Chen G, Wang X, et al. Environmental, energy, and economic analysis of integrated treatment of municipal solid waste and sewage sludge: A case study in China. Science of the Total Environment. 2019;**647**:1433-1443

[12] Ding A, Zhang R, Ngo HH, He X, Ma J, Nan J, et al. Life cycle assessment of sewage sludge treatment and disposal based on nutrient and energy recovery: A review. Science of the Total Environment. 2021;**769**:144451

[13] Hong JL, Hong JM, Otaki M, Jolliet O. Environmental and economic life cycle assessment for sewage sludge treatment processes in Japan. Waste Management. 2009;**29**(Pt. 2):696-703

[14] Lishan X, Tao L, Yin W, Zhilong Y, Jiangfu L. Comparative life cycle assessment of sludge management: A case study of Xiamen, China. Journal of Cleaner Production. 2018;**192**:354-363

[15] Liu F, Liu H, Yang N, Wang L. Comparative study of municipal solid waste incinerator fly ash reutilization in China: Environmental and economic performances. Resources, Conservation and Recycling. 2021;**169**:105541

[16] Di Capua F et al. High-solid anaerobic digestion of sewage sludge: Challenges and opportunities. Applied Energy. 2020;**278**:115608

[17] Di X, Nie Z, Zuo T. Life cycle emission inventories for the fuels consumed by thermal power in China. China Environmental Science. 2005;**25** (Pt. 5):632-635

[18] Luo N. Research on Environmental Impact of Sintered Brick Production in China. China: Beijing University of Technology; 2009. (In Chinese)

[19] Song DL, Hou SP, Wang XB, et al. Nutrient resource quantity of crop straw and its potential of substituting. Journal of Plant Nutrition and Fertilizers. 2018; **24**(Pt. 1):1-21. (In Chinese)

[20] Edwards J, Burn S, Crossin E, Othman M. Life cycle costing of municipal food waste management systems: The effect of environmental externalities and transfer costs using local government case studies. Resources, Conservation and Recycling. 2018;**138**:118-129

[21] Goedkoop M, Heijungs R, et al. Recipe 2008. The Hague, Netherlands: Dutch Ministery of the Environment; 2009

[22] Itsubo N, Sakagami M, Kuriyama K, Inaba A. Statistical analysis for the development of national average weighting factors—Visualization of the variability between each individual's environmental thoughts. The International Journal of Life Cycle Assessment. 2012;**17**(Pt. 4):488-498

[23] Martinez-Sanchez V, Levis JW, et al. Evaluation of externality costs in life-cycle optimization of municipal solid waste management systems. Environmental Science and Technology. 2017;**51**(Pt. 6):3119-3127

[24] Wu X, Zhang Z, Chen Y. Study of the environmental impacts based on the "green tax"—Applied to several types of building materials. Building and Environment. 2005;**40**(Pt. 2):227-237

[25] Zhang J, Liang Z, Matsumoto T, Zhang T. Environmental and economic implication of implementation scale of sewage sludge recycling systems considering carbon trading price. Sustainability. 2022;**14**:8684

[26] Guo D, Li B, Yu W, Baroutian S, Young BR. A system engineering perspective for zero carbon emission in wastewater and sludge treatment: A review. Sustainable Production and Consumption. 2024;**46**:369-381